Lowell

W9-BAQ-381

The Story of an Industrial City

A Guide to
Lowell National Historical Park
and Lowell Heritage State Park
Lowell, Massachusetts

Produced by the
Division of Publications
National Park Service

U.S. Department of the Interior
Washington, D.C.

Using this Handbook

Lowell, Massachusetts was America's first large-scale planned industrial community. After its establishment as a town in 1826, Lowell was celebrated for its innovative textile technology and its unique workforce of young Yankee farm women. Its mills helped transform American life with high-volume mechanized manufacturing, the rise of the corporation, and the growth of an urban working class. Lowell National Historical Park preserves many scenes of the city's industrial past. This book is both an illustrated account of the changing fortunes of Lowell's industry and its working people and a guide to surviving sites.

The publication of this book was supported by grants from the Massachusetts Foundation for Humanities and the Lowell Plan, Inc.

National Park Handbooks help promote understanding and enjoyment of America's natural and historical inheritance. They are sold at parks and can be purchased by mail from the Superintendent of Documents, U.S. Government Printing Office, Washington, D.C. 20402.

Library of Congress Cataloging-in-Publication Data
Lowell: the story of an industrial city: a guide to Lowell National Historical Park and Lowell Heritage State Park, Lowell, Massachusetts/produced by the Division of Publications, National Park Service.
p.   cm. — (Official national park handbook; handbook 140)
Includes bibliographical references.
ISBN 0-912627-46-8
1. Textile industry — Massachusetts — Lowell — History.
2. Textile workers — Massachusetts — Lowell — History.
3. Lowell (Mass.) — Economic conditions. 4. Lowell (Mass.) — Social conditions.
I. United States. National Park Service. Division of Publications.
II. Series: Handbook (United States. National Park Service. Division of Publications); 140.
HD9858.L9L69   1992   974.4'4 — dc20
91-28814   CIP

Contents

The Spirit of the Past

America's self-image is founded in part on the nation's rapid rise to industrial preeminence by World War I. While there is no single birthplace of industry, Lowell's planned textile mill city, in scale, technological innovation, and development of an urban working class, marked the beginning of the industrial transformation of America. Visitors can see today the working components of this early manufacturing center—the dam and nearly six miles of canals that harnessed the energy of the Merrimack River; the mills where the cloth was produced; a boardinghouse representing the dozens of like buildings that housed the workers; the churches where they practiced their faiths; the ethnic neighborhoods. These are the roots of American industry and of American working people.

The Lowell story is as much about change as about beginnings. Just as the city today reflects the deindustrialization happening across our northern states, so its historical structures represent one of the greatest transitions in American social history. This was the shift from a rural society, where most people adapted their lives to natural cycles, to a society in which people responded to factory bells, where work was the same year round and did not cease at nightfall. In these pages historian Thomas Dublin tells of the changes undergone by Lowell: the city's role in the Industrial Revolution; the replacement of Yankee women on the mill floor by immigrant men and women; the transition from waterpower to steam; Lowell's decline following the shift of textile capital to the South.

Nowhere is Lowell's transformation more vividly seen than in its workforce. Successive waves of immigrants came to Lowell, taking the lowest paying jobs as those who had come before climbed the economic ladder. One of the most moving of the city's monuments is the group of 20 bronze bricks laid in the sidewalk that leads to Boott Mills. Inscribed on them are names like True W. Brown, Patrick Kelley, Charles Demers, Karoline Alrzybala—the kinds of Yankee, Irish, French-Canadian, and Polish names still found in Lowell. In this unobtrusive setting is distilled the spirit of the city. Its people remain the vital connection to the generations who labored here.

All of Lowell's grand scale, enormous power, and advanced engineering could be brought to bear on the making of cloth only in partnership with the human hand.
Facing page: *Each spring the rising waters of the Merrimack River cascade over the Pawtucket Dam. Built in 1826, the dam still feeds the city's working canals.*
Preceding pages: *Fitz Hugh Lane painted some of the earliest views of Lowell in its prime. His "Middlesex Company Woolen Mills," done in the late 1830s or early 1840s, shows female workers entering the building between pieces of cloth drying on tenterhooks.*
Pages 8-9: *Although some of the structures that made up Lowell's "mile of mills" are gone, the walls still looming over the Merrimack River hint at how powerful a presence they must have been in the surrounding 19th-century countryside.*
Pages 10-11: *Turbine-driven pulleys and belts in the Pawtucket Gatehouse open ten sluice gates that control the flow of water from the Merrimack into the Northern Canal.*

7

Part 1

Beginnings of Industrial America

Seeds of Industry

The rise of Lowell in the second quarter of the 19th century prompted flights of rhetoric from poets and politicians. Massachusetts Governor Edward Everett wrote that the city's tremendous growth "seems more the work of enchantment than the regular process of human agency." John Greenleaf Whittier mirrored these sentiments. Lowell was "a city springing up," he said, "like the enchanted palaces of the Arabian Tales, as it were in a single night—stretching far and wide its chaos of brick masonry. . . . [The observer] feels himself . . . thrust forward into a new century."

The city was an obligatory stop for Europeans touring the United States. To French political economist Michel Chevalier, sent to the United States in 1834 to study American industry, Lowell evoked memories of the Old World before the rise of its infamous industrial cities: "This then is not Manchester . . . Lowell, with its steeple-crowned factories, resembles a Spanish town with its convents, but with this difference, that in Lowell, you meet no rags nor Madonnas, and that the nuns in Lowell, instead of working sacred hearts, spin and weave cotton." The city was sometimes described as one of the wonders of the world. "Niagara and Lowell are the two objects I will longest remember in my American journey," said a Scottish visitor, "the one the glory of American scenery, the other of American industry."

Most visitors were impressed by the sheer scale of mid-19th century Lowell, something best appreciated from across the Merrimack River. Massive five- and six-story brick mills lined the river for nearly a mile, standing out dramatically amid the area's scattered farms. The city itself was only a backdrop; the textile mills dominated the Lowell scene.

Next to the mills, it was the complex network of power canals that caught the eye of visitors to Lowell. By 1850 almost six miles of canals coursed through the city. Operating on two levels, they drove the waterwheels of 40 mill buildings, powering 320,000 spindles and almost 10,000 looms and giving employment to more than 10,000 workers.

Despite the European response to these marvels, we cannot easily contrast "new world" Lowell and "old world" industrial cities. Though Lowell was in many ways new compared to English manufacturing centers, the mills were the product of technological

The centers of New England textile production before Lowell were the Blackstone Valley in Rhode Island and Waltham, Massachusetts. Pawtucket, R.I., was the site of Slater's Mill, the first powered spinning mill.
Facing page: Early folk painting, c. 1814, depicts rural dwellings and a small factory on the Concord River in East Chelmsford, Massachusetts. This mostly agrarian community was transformed into the textile center later renamed Lowell.
Preceding pages: In Winslow Homer's "Morning Bell," young women begin their day's labor at a small rural mill like those in East Chelmsford.

Before industrial Lowell, before rural Chelmsford, Algonquian-speaking Pennacook Indians came to the Pawtucket Falls to take fish from the Merrimack River. Theodore de Bry's engraving of a John White painting (1585) shows southern coastal Indians using similar fishing methods.

and economic developments rooted in 18th-century Europe. The quickening influence of the English Industrial Revolution and the disruption of trade during the Napoleonic Wars helped push America into its own industrial age.

The city's brick mills and canal network were, however, signs of a new human domination of nature in America. Urban Lowell contrasted starkly with the farms and villages in which the vast majority of Americans lived and worked in the early 19th century. Farming was largely a matter of accommodation to the natural world. Mill owners prospered by regimenting that world. They imposed a regularity on the workday radically different from the normal routine. Mills ran an average of 12 hours a day, 6 days a week, for more than 300 days a year. Only when it suited them did the owners follow seasonal rhythms, operating the mills longer in summer but in winter extending the day with whale-oil lamps.

Lowell's canals depended on water drawn from a river, but to use the Merrimack as efficiently as possible, the mill owners dammed it, even ponding water overnight for use the next day. Anticipating seasonal dry spells, they turned the river's watershed into a giant millpond. They were aggressive in purchasing water rights in New Hampshire, storing water in lakes in the spring and releasing it into the Merrimack in the summer and fall.

Damming alone would not have created enough power to run the mills. Lowell's industrial life was sustained by naturally falling water. At Pawtucket Falls, just above the Merrimack's junction with the Concord, the river drops more than 30 feet in less than a mile—a continuous surge of kinetic energy from which the mills harnessed over 10,000 horsepower. Without the falls, there would have been no textile production, no Lowell.

Pawtucket Falls had long been the focus of human activity in the area. If the tumbling water meant power to European settlers, to the nearby Pennacook Indians it was a source of food. Neighboring tribes regularly met at the falls in the spring to reap the bounty of the annual runs of salmon and sturgeon. While Indians planted crops near their villages, they did not "possess" the land or own it individually as the English did. They moved about with the seasons, leaving themselves open to encroachment by settlers who coveted their land. With the incorporation of Chelmsford in 1655, a permanent English presence was established near the Pennacook villages. Conflict and displacement soon followed.

When King Philip's War broke out along the New England frontier in 1675, most Pennacooks followed their sachem Wannalancit north into the New Hampshire woods to avoid hostilities. After their victory, the colonists forced the Indians still living in eastern Massachusetts to move to a few permanent villages, including Wamesit in what is now downtown Lowell. Even so, settlers continued to encroach upon Pennacook lands, and in 1686, Wannalancit formally sold his tribe's rights to land along the Merrimack and Concord Rivers. The remaining Pennacooks moved on to New Hampshire or Canada, and their former lands were absorbed into Chelmsford.

Chelmsford grew steadily throughout the 18th century. At first, ample land and opportunity allowed most sons and daughters to marry and settle within the community. After 1725, however, when the last communal land was distributed, growing numbers of young people seeking farmland had to leave town. The press of population on resources encouraged the remaining villagers to look for work other than farming.

The obvious alternatives, aside from going to sea, were industry and trade. But opportunities there were hardly better. England's mercantile policies generally limited her colonies to such local industries as sawmills and grist mills or crafts like harness making or coppersmithing. There were more people than jobs in these trades. Things were no better after independence. Although the restrictions were gone, American manufacturers found it difficult to compete with the cheap, well-made British goods flooding the market.

It was shipbuilding and its need for timber that transformed this farm community at the end of the 18th century. As a neutral power during the Napoleonic wars, the United States dominated the Atlantic trade for a few years. The demand for new American ships spurred merchants and shipbuilders in Newburyport, a seaport north of Boston, to tap the inland forests for wood. The port was at the mouth of the Merrimack, and timber could be floated downriver. There was but one major obsta-

Newburyport, a shipbuilding center at the mouth of the Merrimack, needed New Hampshire timber. Pawtucket Falls at East Chelmsford was the only obstacle to rafting it down the river, so the seaport's merchants built the Pawtucket Canal to skirt the falls. It later became the spine of Lowell's canal system.

cle, Pawtucket Falls at East Chelmsford, as the section of Chelmsford at the junction of the Merrimack and Concord Rivers was called.

So it was in 1796 that Newburyport merchants built the Pawtucket Canal—1½ miles long with four sets of locks—to circumvent the falls. Eight years later, a much more ambitious project, the Middlesex Canal, opened a direct water route to the port of Boston. This canal began at Middlesex Village, just upstream from East Chelmsford. Horse-drawn barges moved cargo through a system of locks and aqueducts between the Merrimack and Mystic Rivers, a distance of 27 miles.

After the opening of the Pawtucket and Middlesex canals, manufacturing took root in East Chelmsford. Sawmills and a glassworks were built near the canals. In 1801, one Moses Hale added picking and carding machines to his "fulling" mill (where woolen cloth underwent the final steps of shrinking and thickening). His new operations prepared raw wool for spinning and weaving by local farm families. Since his mill was already finishing the woven cloth, Hale's operation was well integrated into the surrounding economy.

These domestic developments were reinforced by the frequent interruptions in trade caused by the Napoleonic Wars and the Embargo of 1807, which cut off American trade with Europe. Unable to import manufactured goods, Americans soon began to supply their own needs, cloth foremost. Moses Hale used this opportunity to expand his business, while two other locals—John Goulding and Jonathan Knowles—built a cotton mill on the Concord River.

The boom in American manufacturing lasted until the end of the War of 1812, when the return of British textiles to the American market drove small manufacturers like Goulding and Knowles out of business. Others, though, saw prospects for profit. Thomas Hurd bought the old Goulding and Knowles mill in 1818, installed power looms, and converted it to woolen manufacture. Moses Hale and Oliver Whipple began to manufacture gunpowder, driving their machinery with water from a new canal fed by the Concord River.

East Chelmsford's shift from agriculture to industry was typical of many New England towns. The transformation began several decades earlier, when

A water-driven saw was among the early applications of the power of the Pawtucket Falls. After the Pawtucket Canal was completed in 1796, sawmills along the canal tapped its water to turn their waterwheels. **Below:** *Alvan Fisher's 1833 painting of Pawtucket Dam under construction records the transformation of Lowell's early landscape.*

When an anti-slavery speaker came to Lowell in 1834, he drew an angry stone-throwing mob. Mill owners and workers depended on Southern cotton, and anyone who threatened the system was unwelcome. Ever since Slater's cotton mill was established in 1790 and the cotton gin invented three years later, Southern cotton and Northern textiles had had a reciprocal relationship. The North's appetite for raw cotton spurred increased cotton production and the expansion of slavery. Lowell not only bought Southern cotton, but it made "negro cloth" that was sold to plantations. For a few years, the machine shop produced cotton gins sold in the South. Senator Charles Sumner called it an "unholy union . . . between the cotton planters and fleshmongers of Louisiana and Mississippi and the cotton spinners and traffickers of New England—between the lords of the lash and the lords of the loom."

Dependence on slave-grown cotton and moral indignation over slavery coexisted uneasily in Lowell in the years before the Civil War. Many Lowell residents were uncomfortable enough about slavery that they opposed its extension into western territories. Most, however, fearing the mounting sectional conflict, probably would have supported a compromise accepting slavery where it already existed. But two years after war broke out, the Union cause and the abolitionists' cause merged. Its "Southern connection" broken, the city lined up behind Lincoln's Emancipation Proclamation and the Union war effort.

Slaves pick cotton under eyes of drivers.

Cotton is ginned before baling.

Bales await shipment in New Orleans.

the nation's first permanent water-powered textile spinning factory was founded in 1790 along the banks of the Blackstone River in Pawtucket, Rhode Island. There the English immigrant Samuel Slater reproduced an Arkwright spinning frame, the first fully power-driven machine for spinning yarn. The partnership of Almy, Brown, and Slater pioneered in carding and spinning cotton by power machinery. Others followed their lead and by 1810 dozens of water-powered spinning mills dotted riverbanks in southern New England.

The spinning mills were small rural affairs and employed few workers. Far more workers earned a living in their own homes by weaving machine-spun cotton yarn on handlooms. If they lived nearby, they picked up yarn at the mills and returned with woven cloth. Those farther away got their yarn from store-keepers who took it on consignment from the mills. This system of outwork helped stabilize the region's economy but was nevertheless an inefficient way of working. Farmers' wives and daughters took their time at weaving, fitting the new work into the routines of farm life. This system was too relaxed for mill owners, who had schedules to meet and orders to fill. Thomas Hurd's decision to install power looms in his mills on the Concord River was typical of the trend of manufacturers to bring all textile operations under their direct control.

Late in 1821 new actors entered the East Chelmsford scene. The small group of men seen looking over the area became the principals in East Chelmsford's transformation from farm village to industrial city. In the next few years they would complete the revolution in American textile production begun by Samuel Slater. They had already taken a bold step in Waltham, on the Charles River to the south. Their achievements foreshadowed what would happen in Lowell on a grander scale, raising it from obscurity to prominence in the industrial history of the nation.

British historian Eric Hobsbawm sharply characterized English industrial history: "Whoever says Industrial Revolution says cotton." Rapid industrialization transformed the lives of English men and women after 1750, and changes in cotton textiles were at the heart of this process.

The manufacture and export of various cloths were vital to the English economy in the 17th and early 18th centuries. Before the Industrial Revolution, textiles were produced under the putting-out system, in which merchant clothiers had their work done in the homes of artisans or farming families. Production was limited by reliance on the spinning wheel and the hand loom; increases in output required more hand workers at each stage.

Invention dramatically changed the nature of textile work. The flying shuttle, patented by John Kay in 1733, increased the output of each weaver and led to increased demand for yarn. This prompted efforts by others to mechanize the spinning of yarn. The first advance came in 1767, when James Hargreaves invented the spinning jenny, allowing one spinner to produce several yarns at a time. Two years later Richard Arkwright patented the water frame, a spinning machine that produced a coarse, twisted yarn and could be powered by water. Coupled with the carding machine, the Arkwright spin-

The Spinning Jenny

Sir Richard Arkwright (1732-92), by combining in one water-powered machine two earlier inventions, produced the first spinning frame to mechanize all the processes that had been performed by hand on a spinning wheel. His "water frame" **(right)***, patented in 1769, was a significant step toward the industrialization of England.*

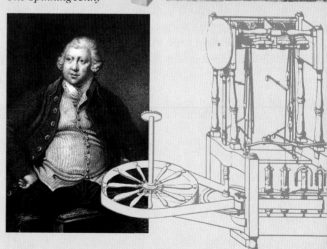

ning frame ushered in the modern factory.

The first textile mills, needing waterpower to drive their machinery, were built on fast-moving streams in rural England. After the 1780s, with the application of steam power, mills also grew up in urban centers. Initially, English mills relied on pauper labor, and for a considerable period mill owners had difficulty recruiting workers. Once in the mills, though, workers felt threatened by the introduction of new machinery, and periodically resisted such moves by destroying power looms and setting fire to new factories. Nevertheless, the textile industry expanded rapidly, increasing production fifty-fold between 1780 and 1840.

The English Industrial Revolution had important consequences for Americans. It spurred cultivation of cotton in the South to meet expanding English demand for the fiber. The growth and profits of English textiles also caught the imagination of American merchants, the more far-sighted of whom sought to manufacture cloth and not simply market English imports. But the degraded conditions and social unrest in English mill towns made many Americans wary of manufacturing. The formidable challenge was to import the innovations without bringing social ills with them.

The Arkwright water frame was best at producing the strong warp yarn that ran the length of the loom. Samuel Crompton's "spinning mule" (above) *combined features of the spinning jenny and the water frame to produce the finer and softer yarns used as weft on the loom. The machines drove a new industrial world where people had to keep up with the quickened pace of automation and where child labor and corporal punishment* (right) *were common practices.*

The mounting conflict between the colonies and England in the 1760s and 1770s reinforced a growing conviction that Americans should be less dependent on their mother country for manufactures. Spinning bees and bounties encouraged the manufacture of homespun cloth as a substitute for English imports. But manufacturing of cloth outside the household was associated with relief of the poor. In Boston and Philadelphia, Houses of Industry employed poor families at spinning for their daily bread. Such practices made many pre-Revolutionary Americans dubious about manufacturing.

After independence there were a number of unsuccessful attempts to establish textile factories. Americans needed access to the British industrial innovations, but England had passed laws forbidding the export of machinery or the emigration of those who could operate it. Nevertheless it was an English immigrant, Samuel Slater, who finally introduced British cotton technology to America. Slater had worked his way up from apprentice to overseer in an English factory using the Arkwright system. Drawn by American bounties for the introduction of textile technology, he passed as a farmer and sailed for America with details of the Arkwright water frame committed to memory. In December 1790, working for mill owner Moses Brown, he started up the first permanent American cotton spinning mill in Pawtucket, Rhode Island. Employing a

Samuel Slater (1768-1835) went to work at 14 for Jedediah Strutt, former partner of Richard Arkwright. A year after his arrival in America, Slater's apprenticeship bore fruit in Pawtucket, Rhode Island. Justly celebrated for introducing the spinning frame (above) to America, he found the development of an efficient carding machine the greater challenge. The mill Slater designed around his new machines in 1793 (left) was an American adaptation of the traditional English mill.

workforce of nine children between the ages of 7 and 12, Slater successfully mechanized the carding and spinning processes.

By 1800 the mill employed more than 100 workers. A decade later 61 cotton mills turning more than 31,000 spindles were operating in the United States, with Rhode Island and the Philadelphia region the main manufacturing centers. The textile industry was established, although factory operations were limited to carding and spinning. It re-mained for Francis Cabot Lowell to introduce a workable power loom and the integrated factory, in which all textile production steps take place under one roof.

As textile mills proliferated after the turn of the century, a national debate arose over the place of manufacturing in American society. Thomas Jefferson spoke for those supporting the "yeoman ideal" of a rural Republic, at whose heart was the independent, democratic farmer. He questioned the spread of factories, worrying about factory workers' loss of economic independence. Alexander Hamilton led those who promoted manufacturing and saw prosperity growing out of industrial development. The debate, largely philosophical in the 1790s, grew more urgent after 1830 as textile factories multiplied and increasing numbers of Americans worked in them.

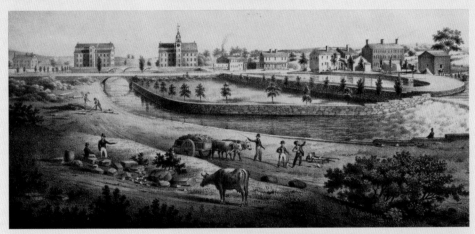

A generation of millwrights and textile workers trained under Slater was the catalyst for the rapid proliferation of textile mills in the early 19th century. From Slater's first mill (seen at right beyond the bridge in center background), the industry spread across New England to places like North Uxbridge, Massachusetts (above). *For two decades, before Lowell mills and those modeled after them offered competition, the "Rhode Island System" of small, rural spinning mills set the tone for early industrialization.*

The Pawtucket and Middlesex Transportation Canals

Between 1790 and 1860 America underwent a transportation revolution. Canals, turnpikes, and railroads criss-crossed the nation, dramatically improving inland transportation.

Eastern Massachusetts was an early participant in this revolution. The first effort to improve navigation on the Merrimack River came in 1792 when a group of investors from Newburyport—some of the same families that later invested in Lowell—chartered the Proprietors of Locks and Canals on Merrimack River. They intended to construct a canal around Pawtucket Falls at East Chelmsford and thereby connect New Hampshire directly to Newburyport at the river's mouth. By 1796 the 1½-mile Pawtucket Canal permitted log rafts and limestone-bearing barges to skirt the falls at the future site of Lowell.

Even while the Pawtucket Canal was under construction, Boston entrepreneurs undertook another ambitious canal venture. Completed in 1803, the Middlesex Canal used 20 locks and 7 aqueducts, over a length of 27 miles, to join the Merrimack River a mile above Pawtucket Falls to the port of Boston.

Together these two canals provided excellent transportation. The Middlesex, however, soon became the more prosperous canal, at least until the industrialists from Waltham discovered the Pawtucket's waterpower potential.

"A Plan of Sundry Farms &c. at Patucket in the Town of Chelmsford," 1821, shows Pawtucket Falls, upper left, and Pawtucket Canal, bottom.

Ten years after the owners of the Middlesex Canal signed an agreement **(left)** to capitalize a canal from the Merrimack River to Boston, the canal, pictured in these scenes, was bringing the products of northern Massachusetts and New Hampshire to the port city. The industrialists from Waltham were drawn to East Chelmsford, later renamed Lowell, because of the falls and the nearby canal. From the start, when the bricks for the first mills were brought in by canal, Lowell relied heavily on both the Middlesex and the Pawtucket. The Middlesex Canal fought burrowing muskrats ("musquashes") and minks by posting bounties, only to speed its own demise by carrying the crossties for the Boston and Lowell Railroad completed in 1835.

Picking removed foreign matter (dirt, insects, leaves, seeds) from the fiber. Early pickers beat the fibers to loosen them and removed debris by hand. Machines used rotating teeth to do the job, producing a thin "lap" ready for carding.

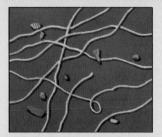

Carding combed the fibers to align and join them into a loose rope called a "sliver." Hand carders pulled the fibers between wire teeth set in boards. Machines did the same thing with rotating cylinders. Slivers were then combined, twisted, and drawn out into "roving."

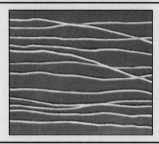

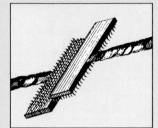

Spinning twisted and drew out the roving and wound the resulting yarn on a bobbin. A spinning wheel operator drew out the cotton by hand. A series of rollers accomplished this on machines called "throstles" and "spinning mules."

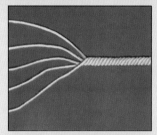

Warping gathered yarns from a number of bobbins and wound them close together on a reel or spool. From there they were transferred to a warp beam, which was then mounted on a loom. Warp threads were those that ran lengthwise on the loom.

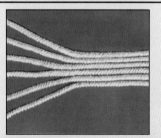

Weaving was the final stage in making cloth. Crosswise woof threads were interwoven with warp threads on a loom (see pp. 48-49). A 19th-century power loom worked essentially like a hand loom, except that its actions were mechanized.

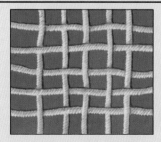

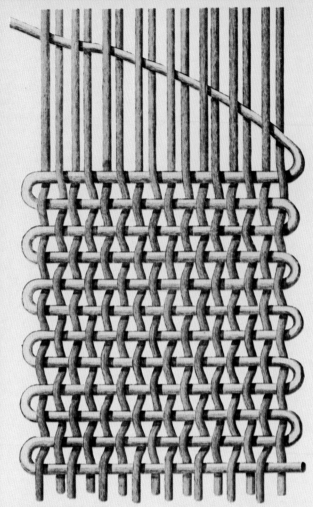

Like food and shelter, clothing is a basic human requirement. When settled neolithic cultures discovered the advantages of woven fibers over animal hides, the making of cloth, drawing on basketry techniques, emerged as one of humankind's fundamental technologies. From the earliest hand-held spindle and distaff and basic hand loom to the highly automated spinning machines and power looms of today, the principles of turning vegetable fiber into cloth have remained constant. Plants are cultivated and the fiber harvested. The fibers are cleaned and aligned, then spun into yarn or thread. Finally the yarns are interwoven to produce cloth. Today we also spin complex synthetic fibers, but they are still woven together the way cotton and flax were millennia ago.

City on the Merrimack

Thousands of female workers, like the one in Winslow Homer's portrayal of a young woman tending a loom, labored in the mills of Lowell. This woodcut illustrates lines from William Cullen Bryant's poem, "The Song of the Sower": . . . for those who throw/The Clanking shuttle to and fro,/In the long row of humming looms. . . .
Facing page: *Although by 1830 Lowell was a busy textile center, there were still quiet, almost rural, parts of town, as shown in this 1834 engraving.*

On a snowy November day in 1821, a small party of men walked through East Chelmsford. "We perambulated the grounds, and scanned the capabilities of the place," Nathan Appleton later reminisced, "and the remark was made that some of us might live to see the place contain twenty thousand inhabitants." Appleton and Patrick Tracy Jackson, another member of the group, were Boston merchants turned industrialists. Seven years earlier they had been principals in the development of cotton mills in Waltham. Now they wanted to make East Chelmsford a textile center.

The man most responsible for their success in Waltham—and for whom East Chelmsford would be renamed—was absent. Francis Cabot Lowell had died before his colleagues began planning an industrial city of unprecedented order and scale. Lowell was a product of the prosperous merchant shipping class of Newburyport and Boston. Cabots, Lowells, Jacksons, Duttons, and Wendells—they all intermarried and formed lifelong business relationships. It was a close-knit system that kept their class dominant in New England for generations. Francis Lowell prospered in this setting but grew weary of the constant risks that were the merchant's lot. As an importer of English textiles, he was aware of the growing demand for cloth and of the infant textile industry's potential as a more secure investment. Yet he was skeptical of the factory system and the horrors it nurtured in English industrial cities.

When Lowell took his family to Great Britain in 1810 to recover his health, he used the occasion to tour the textile mills of Manchester, and some say to engage in industrial espionage. He apparently memorized the workings of the power looms common in these mills. He may also have visited a few of the Scottish "improvements," planned villages established by enlightened landowners so their tenants could live and work in decent, healthy surroundings. The sense of order would have appealed to Lowell. He saw rapid industrial change coming to America, and he hoped to profit from it while holding on to the stable society to which he and others of his class were accustomed.

Returning home two years later, Lowell set out to reproduce the type of power loom he had seen, and with it place American textile manufacturing on a

Lowell's city seal used "art" in its broad sense to mean the human ability to make or create—in Lowell's case, millions of yards of cloth.

new footing. After Lowell gained the confidence of his brother-in-law Patrick Jackson, they pooled their capital and solicited other investors. They called their enterprise the Boston Manufacturing Company and built their first mill in Waltham, a few miles up the Charles River from Boston. By 1814, Lowell and Paul Moody, an inventive mechanic, had developed a power loom partly based on what he had seen in England.

Lowell succeeded in devising a fully integrated textile mill at Waltham. For the first time in America, all steps in the textile manufacturing process—from bale to bolt—were conducted in one mill, an innovation that became known as the "Waltham-Lowell system."

The Boston Manufacturing Company enjoyed remarkable success in its early years. The company concentrated on coarse cloth for the growing domestic market, and its monopoly of power weaving gave it a significant advantage over competitors. The investors began planning to expand the business, but not at Waltham. The three mills there were using all of the available waterpower. Jackson and Appleton—Lowell had died in 1817—sought a new site with room to grow.

Their search ended at East Chelmsford, a farming village on the Merrimack River 30 miles from Boston. Jackson was familiar with the area and understood the potential of Pawtucket Falls. The existence of the Pawtucket Canal simplified the construction of power canals, and the proximity of the Middlesex Canal assured transportation to a major port. So in the fall of 1821 directors of the Boston Manufacturing Company—the Boston Associates, as they are called by historians—bought up shares in the corporation that operated the Pawtucket Canal and quietly purchased farmland between the canal and river.

Construction at the new site began in 1822. Kirk Boott, an imperious former officer in the British army, took charge. As planner, architect, engineer, and construction boss, Boott oversaw the building of mills, canals, locks, machine shop, and worker housing. He designed buildings and laid out streets. He even planned the first church in Lowell, St. Anne's. Boott decided it would be Episcopalian, his faith, even though most of the workers belonged to other sects.

Among Boott's earliest construction crews were about 30 Irish laborers who walked all the way from Boston to dig canals—forerunners of the waves of newcomers, both Yankee and immigrant, that over the next decades would transform a rural community into a new industrial city. These first Irish made do in the roughest of conditions. Living off wages of 75 cents a day, they crowded into tents and shacks in the "paddy camp lands," later known as the Acre.

In 1823 the Irish completed the first branch of the Pawtucket Canal, the Merrimack Canal. The Merrimack, which carried water to the Merrimack Manufacturing Company mill, took advantage of the entire fall of the river. Boott was impressed by the power of the 30-foot waterwheel, noting in his diary: "Sept 4, 1823. After breakfast, went to factory, and found the great [water] wheel moving around his course, majestically and with comparative silence. Moody [the wheel's designer] declared that it was 'the best wheel in the world.' "

Once the East Chelmsford factory was up and running, the Boston Associates set about expanding their new operation. From the beginning, they saw the potential for profit in real estate and the sale of water rights. They also needed a company to oversee construction and technological development. In 1825 the company directors made a strategic move. They restructured the Proprietors of Locks and Canals Company (the original owner of the old Pawtucket Canal) as a tool to manage these activities. The first accomplishments of the Locks and Canals Company were the completion of the canal system and the permanent damming of the Merrimack River in 1826.

This company's formidable power lay in its control of land and waterpower. It took charge of every aspect of establishing a new textile mill. It sold the land, leased the water rights, put up the mill buildings, outfitted them with machinery crafted in its machine shop, constructed whatever new canals and roads were needed, and furnished housing for the workers. In effect it sold prepackaged mills to new textile firms, which were generally owned by the same investors who controlled the Boston and the Merrimack Manufacturing Companies.

Lowell was an immediate success both as a manufacturing center and as a real estate development. The Merrimack Company, concentrating on printed

The textile machinery at Lowell was modeled on that of England. However, the early mill owners hoped to avoid the social problems that arose from the English factory system—typified by the Lancashire mills shown here. Above all they did not want to create a permanent underclass of discontented factory workers.

The Waltham-Lowell System and the Boston Associates

The success of the early spinning mills of southern New England in the years before 1810 and the uncertainties of shipping led the son of a leading Boston merchant family, Francis Cabot Lowell, to seek a haven for his fortune in manufacturing. Having developed the country's first working power loom, Lowell, with fellow Bostonians Patrick Tracy Jackson and Nathan Appleton, established the Boston Manufacturing Company along the Charles River in Waltham (below) in 1814.

There Lowell and his fellow entrepreneurs, later called the "Boston Associates," transformed the country's fledgling textile industry. Capitalized at $400,000, the Waltham mill dwarfed its competition. The power loom and related machinery permitted the combination of all the steps in the production of cloth under a single roof. Instead of relying on traditional family labor, the company recruited young single women from the surrounding countryside. So great were the profits at Waltham that the Boston Associates soon

Francis Cabot Lowell

Nathan Appleton

Patrick Tracy Jackson

looked for new sites, first at East Chelmsford (renamed Lowell), and then Chicopee, Manchester, and Lawrence. The "Waltham-Lowell system" succeeded beyond their expectations, giving the Boston Associates control of a fifth of America's cotton production by 1850.

Their profits permitted this tight-knit group of families— Appletons, Cabots, Lowells, Lawrences, Jacksons—to build an economic, social, and political empire. They helped develop the Boston and Lowell Railroad and other railroad lines in New England. They owned controlling stock in a host of Boston financial institutions, allowing them to finance and insure ventures through their own companies. As their fortunes grew, the Boston Associates turned to philanthropy—establishing hospitals and schools—and to politics, playing a prominent role in the Whig Party in Massachusetts. Until the Civil War, the Boston Associates were New England's dominant capitalists.

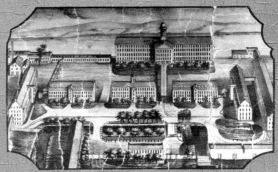

Merrimack Manufacturing Company, 1850

Lowell's machine shop complex was second in importance only to the textile mills among the city's industries. Incorporated as an independent company in 1845, the Lowell Machine Shop had its origins as the machine shop of the Boston Manufacturing Company in Waltham from 1814 to 1824. The Merrimack Company in Lowell then housed the machine shop, which was taken over by the Proprietors of Locks and Canals in 1825.

The shop underlay Lowell's textile industries: fabricating machines that turned cotton into cloth, building waterwheels, turbines, and steam engines that provided the power, and making shafts, gears, and pulleys that transferred power within the mill. Its influence extended beyond Lowell, as it built machine tools and complete sets of machinery for mills in other cities. The locomotives also produced there helped transform New England's transportation system.

Some of the city's best minds headed the machine shop in its early years. Paul Moody, the mechanic who had helped Francis Cabot Lowell develop the power loom at Waltham, was head of Waltham's machine shop. Under Moody, the machine shop was creative and versatile. There the machines for the first mill building in Lowell were built. When the Waltham machinery was moved to Lowell, Moody followed, becoming head of the machine shop upon its establishment there in 1824.

George Washington Whistler directed the building of the shop's first loco-

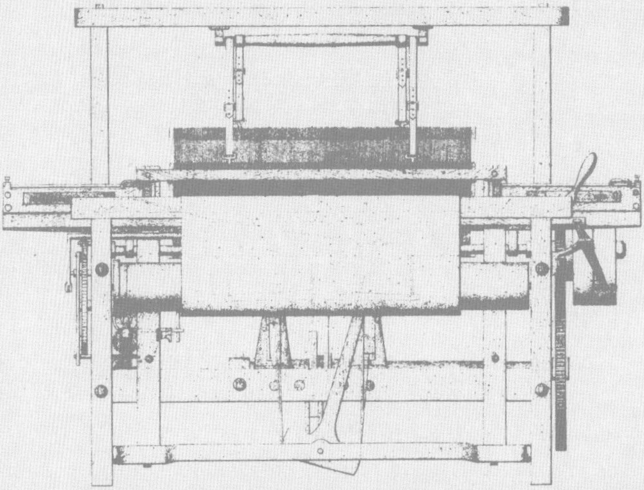

The power loom (1848 illustration at left) developed by Francis Cabot Lowell and Paul Moody became a staple of the machine shop. Many of the thousands produced in Lowell went to mills in other textile cities.

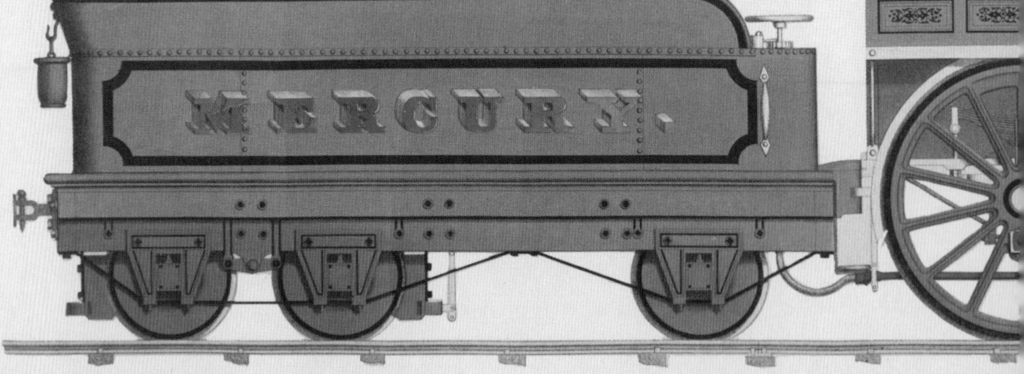

motive in 1835. He took apart an English locomotive imported from the Stephenson works at Newcastle to learn how it was constructed. From the components, Whistler fabricated patterns from which the shop manufactured its own locomotive — one of New England's earliest. Three years later the shop had turned out 32 locomotives.

James B. Francis, who took charge of the machine shop in 1837, was the major figure in Lowell's engineering history. He fine-tuned the city's canal system, engineered the Northern Canal, and oversaw Lowell's transition to turbines. Under Francis' direction, the Lowell Machine Shop became a leader in the fabrication of hydraulic turbines.

The development of such skills in the textile industry's early machine shops was a crucial step in the American Industrial Revolution. Previously, Americans relied heavily on English expertise and machines. It took fine tools to make tools, precise machines to make other machines. The process was slow, and required patient trial and error and borrowed technology before Americans learned to make their own. Much of this learning took place in the Waltham and Lowell shops, where Paul Moody helped train the first generation of master mechanics. Out of these and others' efforts to emulate British textile technology came the machine tool industry on which other industries were founded. It was the beginning of America's machine age.

George Washington Whistler (1800-1848), a prominent engineer who built railroads in America and Russia, was brought to Lowell to develop the machine shop's locomotive capacity. **Below:** *1852 lithograph advertised one of the types of locomotives built by the machine shop.*

Kirk Boott (1790-1837) was a central figure in Lowell's early history. As first agent of the Merrimack Company, he laid out streets, designed mills, oversaw construction, and established rules for workers. Alternately praised as a commanding leader and damned as an autocrat, he put his stamp on the city.
Facing page: *Employees in the Boott Mills courtyard, c. 1880.*

calicos, was highly profitable. New textile companies sprang up in Lowell in rapid succession—Hamilton (1826), Appleton (1828), Lowell (1829), Middlesex (1831), Suffolk and Tremont (1832), Lawrence (1833), Boott (1836), and Massachusetts (1840)—filling up all power sites along the canals. This expansion reflected strong returns for the textile firms and yielded enormous profits—averaging 24 percent a year between 1824 and 1845—for Locks and Canals. By 1846 the mills were turning out almost one million yards of cloth a week, and by 1850 there were 10 large mill complexes employing more than 10,000 people. Until the Civil War, Lowell was the largest concentration of industry in America.

The booming population of the American West and South absorbed much of the city's textile goods, but the mill selling agents aggressively sought out new foreign markets in South America, China, India, Russia, and other parts of the world. The quality and low price of their cloth often enabled the Lowell companies to outstrip English and other American producers in these markets.

The population of Lowell grew dramatically during the years of rapid industrial expansion—rising from about 2,500 in 1826, to 16,000 in 1836, to more than 33,000 by 1850, when Lowell was the second largest city in Massachusetts. Other New England mill towns emerged, often modelled after Lowell and established by the same group of investors who had transformed East Chelmsford. But in the early years Lowell kept a step ahead.

The Boston Associates benefited from a good reputation in these first decades. In their public pronouncements they emphasized the good working conditions, the short-term nature of employment, and the quality of their workers, arguing that Lowell's industrial system was consistent with republican values. But their central motive was profits rather than social uplift, and the owners did their best to minimize competition among the companies they controlled. Lowell firms commonly shared officers and boards of directors. It was typical of Lowell mill agents to set wages at common levels, to share data on the costs of production, and to enforce a blacklist of dismissed workers. These interlocking controls were an essential ingredient in the Waltham-Lowell system.

When Chief Engineer James B. Francis erected a flood gate at Guard Locks on the Pawtucket Canal, people ridiculed his effort, calling it "Francis' Folly." Two years after the gate's completion in 1850, the Merrimack rose 14 feet over Pawtucket Dam. Francis' high wooden gate was dropped across the lock, holding back the water and saving the city from flooding.
Facing page: *Workers' days were regulated from the first pre-dawn bell until the evening bell some 14 hours later. In 1853 the workday was reduced to 11 hours.*

What were the system's other characteristics? Most important, it combined all steps in cloth manufacturing within one mill. Factories were no longer limited to carding and spinning or dependent on dispersed farm families for weaving yarn into cloth. Virtually all steps from opening bales of cotton to bleaching or printing the cloth were mechanized. There was also a new scale to the operations. By the mid-1830s, for instance, the Merrimack Company employed roughly a thousand workers, 20 times the number found in typical manufacturing firms elsewhere.

The Lowell mills differed from other industrial enterprises in another way. Their workers were mostly young farm women recruited from the surrounding countryside. In contrast, earlier textile mills in southern New England had employed whole families, and urban workshops generally hired skilled male artisans. To house these women, the companies built scores of boardinghouses. In the mid-1830s nearly three-fourths of the female workers lived in boardinghouses, usually under the charge of responsible older women. The rest lived at home with their parents or with other relatives in Lowell.

Yet another feature that distinguished the Lowell mills in this period was the monthly payment of cash wages. Most other employers paid workers with credit at a company store or settled wages four times a year. In Lowell during the 1830s, a woman might earn $12 to $14 a month. After paying $5 monthly for room and board in a company boardinghouse, she had the rest for clothing, tickets to lectures, savings, or incidentals. She could never have earned this much money at farm work and quite likely had more ready cash than her father. It was common for young women to return home after a year in the mills with $25 to $50 in a bank account.

Wages were not Lowell's sole attraction for women. The city also offered social, cultural, and religious opportunities. In evenings after work, the women might attend a lecture, exhibition, or play. They could subscribe to magazines and newspapers that were probably unknown in the countryside. Some joined lending libraries or literary circles that offered intellectual stimulation. The city's clothing and dry goods stores put those in their home towns to shame. A wide array of Protestant churches offered sabbath services, Sunday schools, and various social activi-

continued on page 50

40

TIME TABLE OF THE LOWELL MILLS,

To take effect on and after Oct. 21st, 1851.

The Standard time being that of the meridian of Lowell, as shown by the regulator clock of JOSEPH RAYNES, 43 Central Street

	From 1st to 10th inclusive.				From 11th to 20th inclusive.				From 21st to last day of month.			
	1st Bell	2d Bell	3d Bell	Eve.Bell	1st Bell	2d Bell	3d Bell	Eve.Bell	1st Bell	2d Bell	3d Bell	Eve.Bell.
January,	5.00	6.00	6.50	*7.30	5.00	6 00	6.50	*7.30	5.00	6.00	6.50	*7.30
February,	4.30	5.30	6.40	*7.30	4.30	5.30	6.25	*7.30	4.30	5.30	6.15	*7.30
March,	5.40	6.00		*7.30	5.20	5.40		*7.30	5.05	5.25		6.35
April,	4.45	5.05		6.45	4.30	4.50		6.55	4.30	4.50		7.00
May,	4.30	4.50		7.00	4.30	4.50		7.00	4.30	4.50		7 00
June,	"	"		"	"	"		"	"	"		"
July,	"	"		"	"	"		"	"	"		"
August,	"	"		"	"	"		"	"	"		"
September,	4.40	5.00		6.45	4.50	5.10		6.30	5.00	5.20		*7.30
October,	5.10	5.30		*7.30	5.20	5.40		*7.30	5.35	5.55		*7.30
November,	4.30	5.30	6.10	*7.30	4.30	5.30	6.20	*7.30	5.00	6.00	6.35	*7.30
December,	5.00	6.00	6.45	*7.30	5.00	6.00	6.50	*7.30	5.00	6.00	6.50	*7.30

* Excepting on Saturdays from Sept. 21st to March 20th inclusive, when it is rung at 20 minutes after sunset.

YARD GATES,

Will be opened at ringing of last morning bell, of meal bells, and of evening bells; and kept open Ten minutes.

MILL GATES.

Commence hoisting Mill Gates, Two minutes before commencing work.

WORK COMMENCES,

At Ten minutes after last morning bell, and at Ten minutes after bell which "rings in" from Meals.

BREAKFAST BELLS.

During March "Ring out"........at:....7.30 a. m........."Ring in" at 8:05 a. m.
April 1st to Sept. 20th inclusive.....at....7 00 " " .... " " at 7.35 " "
Sept. 21st to Oct. 31st inclusive.....at....7.30 " " .... " " at 8.05 " "
Remainder of year work commences after Breakfast.

DINNER BELLS.

"Ring out"...... ....12.30 p. m........."Ring in".... 1.05 p. m.

In all cases, the *first* stroke of the bell is considered as marking the time.

B. H. Penhallow, Printer, 28 Merrimack Street.

Lowell's Canal System 1850

The Lowell canal system evolved steadily from 1821, when the Boston Associates purchased the old Pawtucket transportation canal in East Chelmsford. They initially used the Pawtucket as a feeder canal to channel water into new power canals. Just above Swamp Locks, the Merrimack, Western, and Hamilton canals branched off, taking water to the Merrimack, Lowell, Tremont, Suffolk, Lawrence, Hamilton, and Appleton mills. Only the Merrimack Company used the full 30-foot drop of water; for other mills the drop was 13 or 17 feet.

In 1847 the construction of the Northern Canal (right) increased waterpower generated by the canal system by 50 percent. By mid-century the canal system we see in Lowell today was in place. Including almost 6 miles of canals and operating on 2 levels, this system powered 10 major mill complexes employing more than 10,000 workers.

PAWTUCKET FALLS

NORTHERN CANAL (1847)

Pawtucket Bridge

Pawtucket Gatehouse

PAWTUCKET DAM
Elevation 101 feet

Ford

Merr

Lowell Street

Street

Fletcher

Bowers Street

School

PAWTUCKET CANAL

Pawtucket

Walker

Street

NORTH COMMON

Street

Cross Street

Street

Lagrange

Francis Gate

Guard Locks

PAWTUCKET CANAL
(converted to power canal, 1823)

(1796)

CANAL LEVELS

Guard Locks

Swamp Locks
13 foot drop

Lower Locks
17 foot drop

30 foot vertical drop

| UPPER CANAL |
| LOWER CANAL |
| LOWER RIVER |

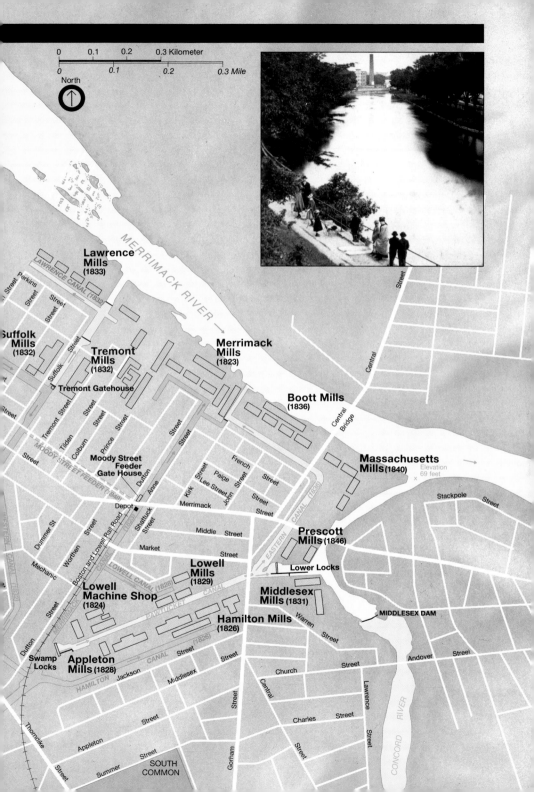

0   0.1   0.2   0.3 Kilometer

0   0.1   0.2   0.3 Mile

North

MERRIMACK RIVER

Lawrence Mills (1833)

LAWRENCE CANAL (1832)

Perkins Street

Street

Street

Street

Suffolk Mills (1832)

Tremont Mills (1832)

Suffolk Street

Tremont Gatehouse

Merrimack Mills (1823)

Street

Street

Tremont Street

Tilden

Coburn

Prince Street

Street

Street

Boott Mills (1836)

Central Bridge

Massachusetts Mills (1840)

Street

Central

Elevation 69 feet

Stackpole   Street

MOODY STREET FEEDER (1826)

Moody Street Feeder Gate House

Dutton

Anne

Kirk Street

Lee Street

Paige Street

John

French Street

Street

Street

Street

EASTERN CANAL (1835)

Prescott Mills (1846)

Depot

Shattuck

Boston and Lowell Rail Road

WESTERN CANAL (1832)

Dummer St

Worthen Street

Mechanic

Market

Merrimack

Middle   Street

Street

Lowell Mills (1829)

LOWELL CANAL (1828)

Lower Locks

Middlesex Mills (1831)

Warren Street

Street

Lowell Machine Shop (1824)

PAWTUCKET CANAL

Hamilton Mills (1826)

Street

MIDDLESEX DAM

Dutton Street

Swamp Locks

MERRIMACK CANAL

Appleton Mills (1828)

HAMILTON CANAL (1826)

Jackson

Middlesex

Street

Church Street

Central Street

Andover   Street

Street

LAWRENCE

CONCORD RIVER

Thorndike

Appleton

Summer   Street

Street

Street

Gorham Street

Street

Charles Street

Street

SOUTH COMMON

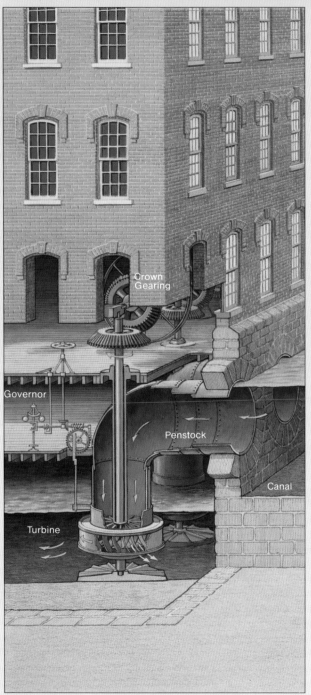

The use of water as a source of motive power dates at least to Greek and Roman civilizations. In ancient times and in the Middle Ages, waterpower seems to have been limited largely to irrigation and the grinding of grains for bread. In colonial America, waterwheels commonly provided power for sawing timber, fulling cloth, grinding grains, and making iron products. Until the second half of the 19th century, waterpower was the major mechanical power source in the United States.

The principles behind waterpower are simple. Basically, a waterpower system taps the potential energy stored in water and turns it into kinetic energy by controlling its natural fall. Water is channeled out of a river at a certain height in a power canal and brought to a point where it is permitted to fall to a lower level. During its fall, it fills the buckets in a waterwheel, its weight driving the wheel around.

Falling water also powers the more efficient turbine, driving it by pressure as well as weight. In the first turbines designed by Uriah Boyden and adapted by James B. Francis to power Lowell's mills, the water entered the wheel at its center and was directed outward by stationary vanes to turn another set of moving vanes. By 1858, 56 Boyden turbines (drawing at right), rated at 35 to 650 horsepower, helped drive Lowell's mills. In both the waterwheel and turbine, the power was transferred by gears to the mill's main power shaft or drive pulley.

Turbine and drive gear arrangement in textile mill.

Breastwheel typical of a rural mill.

Wooden crown gearing off waterwheel

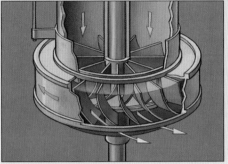

Outward-flow turbine

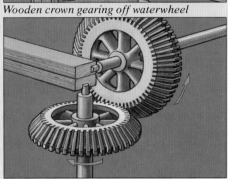

Iron bevel gears off turbine

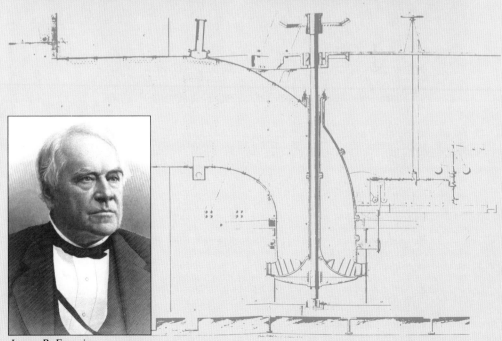

James B. Francis

Once a wheel or turbine had harnessed the water's power, the mill engineer had to transfer the power throughout the mill to hundreds of machines. British and early American mills ran a vertical shaft off the main drive shaft, then transferred the power by gears to overhead shafts on each floor. Because it was difficult to get precisely-machined gears, American mills were rough and noisy and had to be run at slow speeds. A few small mills used belting, but it wasn't until Paul Moody used belt-ing in the Appleton Mills in 1828 that it was seriously considered as an alternative to shafting. Leather belts transferred power directly to the horizontal shafts on each floor. Belts allowed faster speeds and were quieter and less jarring than shafting. Belting was also much lighter, easier to maintain, and more forgiving of imprecise mill construction. By mid-century, belting had become a distinguishing characteristic of American mills.

Paul Moody

Crown Gearing

Flywheel

Turbine

This mill in Manville, R.I., used a typical distribution system, with belts transferring power from turbines and flywheels.

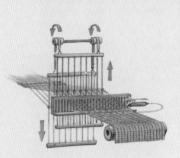

For the first two centuries of American history, the weaving of cloth was a cottage industry, even after the introduction of power spinning frames in 1790. Yarn produced by machines in water-powered factories was still put out for weaving on hand looms in homes. All cloths were woven in basically the same way, although weavers followed patterns to produce cloths with intricate weaves. Because the operations of a loom focus on such a small working area, its movements must be exact. And weaving, as opposed to spinning, requires a cycle of sequential steps and involves reciprocal movement as well as circular. In a power loom, movements coordinated by human hand

Shedding *In the simplest weave pattern, half of the heddles are drawn up, carrying with them alternate warp, or lengthwise, yarns. The other heddles move down. This action separates the yarns into a "shed."*

Picking *The shuttle flies across the shuttle race, carrying the weft, or crosswise, yarn through the shed.*

Warp Thread

Reed

Weft Thread

Beating-In *After the shuttle has made its pass, the reed swings forward to push the newest weft yarn tightly and evenly against the already woven cloth.*

Shed

Heddles

Bobbin

Shuttle Race

Harness

Shuttle

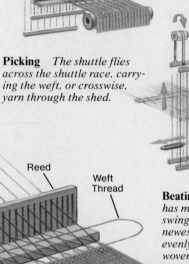

and eye have to be replicated through the precise interaction of levers, cams, gears, and springs. For these reasons, weaving was the last step in textile production to be mechanized.

Successful power looms were in operation in England by the early 1800s, but those made in America were inadequate. Francis Cabot Lowell realized that for the United States to develop a practical power loom, it would have to borrow British technology. While visiting English textile mills, he memorized the workings of their power looms. Upon his return, he recruited master mechanic Paul Moody to help him recreate and develop what he had seen. They succeeded in adapting the British design, and the machine shop established at the Waltham mills by Lowell and Moody continued to make improvements in the loom. With the introduction of a dependable power loom, weaving could keep up with spinning, and the American textile industry was underway.

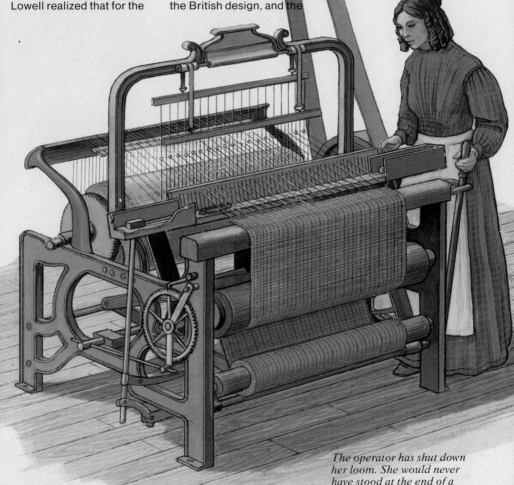

A specially wound and easily replaced bobbin fits in the shuttle (left) and unwinds its yarn smoothly as the shuttle carries it across the loom.

The operator has shut down her loom. She would never have stood at the end of a running machine, as the shuttle sometimes skipped out of the shuttle box and shot through the air, becoming a dangerous missile.

ties. Lowell offered its women workers experiences they could never have known on the farm.

Opportunities like these were dearly won. For their newfound independence, the women were required to stay for at least a year. Their working conditions were hardly healthful. The need to stand all day took its toll. To maintain humidity (necessary to keep the yarn from breaking) the windows were nailed shut, leaving the air filled with lint and making the work rooms hot, damp, and noisy. These conditions left the workers susceptible to lung disease and typhus. The boardinghouses, while certainly an improvement over the living conditions of the typical English textile worker, were crowded and ill-ventilated.

Moreover, the lives of Lowell's female workers were strictly regulated: first, by their hours of work; second, by company rules. In the 1830s the women worked long hours every day except Sunday. The workday was longest in summer, when operatives stayed at their machines for 14 hours, with brief breaks for breakfast and dinner. Hours were shorter in winter. Even then, the operatives worked by oil lamp after sunset, a practice which struck many of

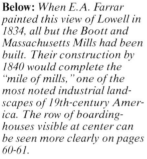

Below: *When E.A. Farrar painted this view of Lowell in 1834, all but the Boott and Massachusetts Mills had been built. Their construction by 1840 would complete the "mile of mills," one of the most noted industrial landscapes of 19th-century America. The row of boardinghouses visible at center can be seen more clearly on pages 60-61.*

the young farm women as contrary to nature. True, farm work was also long and laborious, but the routines of a farm day were looser and more adaptable to human needs.

The factory bells dominated daily life. They woke the workers at 4:30 a.m. on summer mornings, called them into the mills at 4:50, rang them out for breakfast and back in, out and in for dinner, out again at 7 p.m. at the day's close. At 10 the bells rang the curfew. The mills had identical schedules and set their bells to ring in unison. The whole city, it seemed, moved together and did the mills' bidding. To a character in a story in the *Lowell Offering*, a company-subsidized magazine written by the women workers, it was torment: "Up before day, at the clang of the bell—and out of the mill by the clang of the bell—into the mill, and at work, to the obedience of that ding-dong of the bell—just as though we were so many living machines."

The company also regulated the women's conduct outside the mills. For women who did not live with relatives, residence in company boardinghouses was compulsory. Regulations posted by all the firms re-

Young mill workers, c. 1860, pose with shuttles—symbol of textile labor.

quired boardinghouse keepers to report improper conduct to managers. Regular church attendance was expected. For a brief time the Merrimack Company even deducted pew rent from employees' earnings, paying the funds to St. Anne's, the Episcopal church built by Kirk Boott on company land. Though regulation in practice was never quite as harsh as it appeared in print, managers adopted a strict paternalism they assumed was needed to control the newly independent women. They constructed a social system acceptable to a rural public often antagonistic to urban life, thereby protecting Lowell's reputation and assuring a steady stream of new recruits into the mills.

While management sought to control mill operatives, experienced women workers had an important role in socializing newcomers, creating solidarity and a measure of freedom in the workplace. Overseers assigned novices to work alongside old hands. Only after two or three months as a "sparehand" did a new worker begin to work at her own machines. In the boardinghouses, oldtimers poked fun at the rural accents and clothing of women fresh from the farm, and they soon taught the country women the "city way of speaking." Nor did it take long for the newcomers to trade their farm clothing for more stylish urban fashions. Women helped one another adapt to a new way of life, shaping a close-knit community that formed the basis for subsequent labor organization and protest.

Protest came to Lowell in the mid-1830s, only a decade after the first mill opened. Mill management, feeling for the first time the pressure of a competitive market, twice reduced the take-home pay of women workers. Faced with growing inventories and falling prices, owners believed the only way to sustain profits was to cut labor costs. The mill workers, however, were not willing to accept this logic of the marketplace.

In February 1834, after mill agents had announced an upcoming wage reduction, a large group of women threatened to quit the mills if wages were cut. The agents considered it presumptuous, even "bold," for women to make such demands. When one agent dismissed a ringleader, other women followed her out of the mills and marched in a procession about the city. Calling workers out of other

75 Young Women

From 15 to 35 Years of Age,

WANTED TO WORK IN THE

COTTON MILLS!

IN LOWELL AND CHICOPEE, MASS.

I am authorized by the Agents of said Mills to make the following proposition to persons suitable for their work, viz:—They will be paid $1.00 per week, and board, for the first month. It is presumed they will then be able to go to work at job prices. They will be considered as engaged for one year, cases of sickness excepted. I will pay the expenses of those who have not the means to pay for themselves, and the girls will pay it to the Company by their first labor. All that remain in the employ of the Company eighteen months will have the amount of their expenses to the Mills refunded to them. They will be properly cared for in sickness. It is hoped that none will go except those whose circumstances will admit of their staying at least one year. None but active and healthy girls will be engaged for this work as it would not be advisable for either the girls or the Company.

I shall be at the Howard Hotel, Burlington, on Monday, July 25th ; at Farnham's, St. Albans, Tuesday forenoon, 26th, at Keyes's, Swanton, in the afternoon; at the Massachusetts' House, Rouses Point, on Wednesday, the 27th, to engage girls,---such as would like a place in the Mills would do well to improve the present opportunity, as new hands will not be wanted late in the season. I shall start with my Company, for the Mills, on Friday morning, the 29th inst., from Rouses Point, at 6 o'clock. Such as do not have an opportunity to see me at the above places, can take the cars and go with me the same as though I had engaged them.

I will be responsible for the safety of all baggage that is marked in care of I. M. BOYNTON, and delivered to my charge.

I. M. BOYNTON,
Agent for Procuring Help for the Mills.

Many of the young farm women drawn to Lowell by recruiting notices would have been put to work at a spinning frame, such as the one shown below in a manufacturer's advertisement. A rare depiction of one of those women before the advent of photography is the silhouette at left, done in Lowell around 1837.

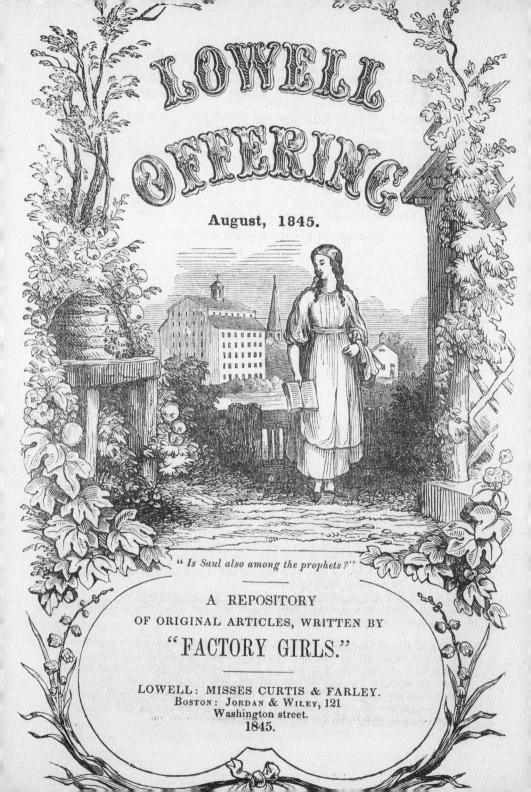

LOWELL OFFERING

August, 1845.

" Is Saul also among the prophets ?"

A REPOSITORY

OF ORIGINAL ARTICLES, WRITTEN BY

"FACTORY GIRLS."

LOWELL: MISSES CURTIS & FARLEY.
BOSTON: JORDAN & WILEY, 121
Washington street.
1845.

Having worked in the Lowell mills from the age of 11, Lucy Larcom (1824-93) moved up to the job of dresser at 16. But the dressing machine, in her words, was a "half-live creature, with its great groaning joints and whizzing fan," that "was aware of my incapacity to manage it, and had a fiendish spite against me." She returned to a lower paying mill job that left her time for literary pursuits. A contributor to the Lowell Offering, *she became the best known of Lowell's celebrated "mill girl" writers.*

mills, the strikers moved to the town common where, according to the Boston *Transcript*, one of them gave a "flaming . . . speech on the rights of women and the iniquities of the 'monied aristocracy.'" Though more than 800 women joined the "turn-out," the mill agents held fast. Within a couple of days most of the strikers returned to work.

Two years later the women mill workers organized a more successful strike. The firms had raised the rates for workers' room and board. In October 1836, about 2,000 women—a third of the female workers in Lowell—joined in the protest. Mill agents reported reduced output for several months running and rescinded the higher charges for many workers.

Taken together, the two turn-outs reveal the Yankee women's pride and independence—qualities that made them sensitive to actions they found exploitative. As "daughters of freemen" steeped in a revolutionary tradition, they were prepared to defend themselves against the harsh economic practices of the mill owners. They did so even though they could be dismissed for striking. Those dismissed had no chance of being employed by other mills, because management circulated the names of troublemaking operatives in a blacklist.

Another period of labor unrest came in the 1840s when Lowell women, joined by other workers across New England, demanded a reduction of hours. Organizing the Lowell Female Labor Reform Association (LFLRA), with mill worker Sarah Bagley as president, they petitioned the state legislature for a 10-hour day. One of the petitions read:

We, the undersigned peaceable, industrious, hardworking men and women of Lowell . . . toiling from thirteen to fourteen hours per day, confined in unhealthy apartments, exposed to the poisonous contagion of air, vegetable, animal, and mineral properties, debarred from proper Physical *exercise, time for* Mental *discipline, and* Mastication *cruelly limited; and thereby hastening us on through pain, disease, and privation, down to a premature grave . . . seek a redress of those evils.*

Several years running, petitions flooded the legislature. The peak came in 1846 when 10,000 workers signed the petitions statewide, more than 4,000 from Lowell alone. When Lowell's representative to the legislature voted with the corporations and opposed the petitions, women mill workers lobbied male voters and helped defeat him at the polls. Yielding to pressure, however, from the textile corporations, the legislature balked at regulating the length of the workday, preferring to let workers and employers settle the issue among themselves. With growing numbers of Irish immigrants ready to replace dissatisfied workers, the owners were increasingly able to ignore their protests. It was not until 1874 that the state enacted a 10-hour law. Even then, manufacturers easily evaded the law.

Other developments in the mills quietly transformed both work and the mill workforce. Responding to competition from new mills—often financed with profits earned in Lowell—the agents required workers to operate additional looms and spindles. They also stepped up the machines' speed: each woman now found herself trying to keep up with more machines, each one running faster. Most despised of the agents' moves was the "premium" system, in which they paid bonuses to overseers who got their workers to produce more cloth than normal.

As some Yankee women responded to these conditions by demanding reduced hours of labor, others voted silently with their feet, leaving the mills. A growing migration from the region to the Midwest left fewer native-born women to replace them on the factory floor. Mill agents were forced to look elsewhere for workers. They found their new labor supply in the Irish immigrants who, after 1845, gradually took the place of Yankee women in Lowell's labor force. The entry of immigrant workers marked the end of an era in Lowell's history and presaged a broader transformation of work and community life in the nation's leading textile city in the second half of the 19th century.

After the Civil War, male mule spinners began to come to the front in labor actions, as women workers had in the 1830s and 40s.

One hundred girls passed through this village on the 29th ult., en route for Lowell; and some fifty for the same destination two weeks since. Agents are sent into this county, Franklin and St. Lawrence, and within the past year, more than four hundred have been "picked up" and forwarded to the factories. Good wages are offered them, or they would not leave their homes, and the great manufacturing establishments are doing a good business that will "pay" or they would not want them.—*Plattsburg Republican.*

As much as the massive brick mills along the Merrimack, "mill girls" were an innovation of the early industrial revolution in New England. Lowell's mill workforce in the antebellum decades consisted largely of young single women from the farming communities of northern New England. Most were between 15 and 25, signing on for short stints

In Lowell's early years, when young New England farm women filled the factories, the textile town still had a decidedly rural look, as seen in the 1825 view above. The earliest surviving photographs of "mill girls" date from decades later. The photo of two female workers was made around 1865. The three Cassidy sisters had their picture taken around 1877.

that rarely exceeded a year at a time. Overall, they averaged about 3 years of employment before leaving the mills for marriage, migration to the west, other employment, or return to their hometowns.

Dissatisfaction with the work environment was a major reason for leaving the mills. In the 1830s and 40s women operatives protested against mill conditions. Their labor movement was not a narrow lobbying effort, but a broad reform campaign embracing a wide range of issues and underpinned by firm ideals. Writing in the *Voice of Industry*, Huldah J. Stone described the attitude of Lowell Female Labor Reform Association members toward reduction of the hours of labor: *They*

do not regard this measure as an end, but only as one step toward the end to be attained. They deeply feel that their work will never be accomplished until slavery and oppression, mental, physical, and religious, shall have been done away with and Christianity in its original simplicity . . . shall be re-established and practiced among men.

In return for monthly cash wages, female workers in Lowell agreed to regulations that varied little from company to company: work for at least a year, live in a company boardinghouse, attend church. Many worked for a year and went back to the farm, some repeating this pattern two or three times.

The Boardinghouse System

The rows of long brick boardinghouses adjacent to Lowell's mills distinguished the city from earlier New England mill towns. Lowell's first female workers at the Merrimack Manufacturing Company were put up in wooden boardinghouses. By the mid-1830s, however, firms were adding brick structures near their mills and requiring women without family in the city to live in them. (A later Merrimack Company boardinghouse is shown below.) Their behavior came under the watchful eye of boardinghouse keepers, who were required to report any misconduct to mill management.

Typically 30 to 40 young women lived together in a boardinghouse. (At right, a group stands in front of what is thought to be a Boott Mills boardinghouse in the 1870s.) The first floor usually contained kitchen, dining room, and the keeper's quarters. Upstairs bedrooms accommodated four to eight women, commonly sleeping two in a double bed. In these close quarters, experienced workers helped new hands adapt to their situation.

The boardinghouses began to fade from prominence as Lowell aged, profits fell, and the workforce changed. At first, Lowell firms rarely accommodated immigrants in the boarding-

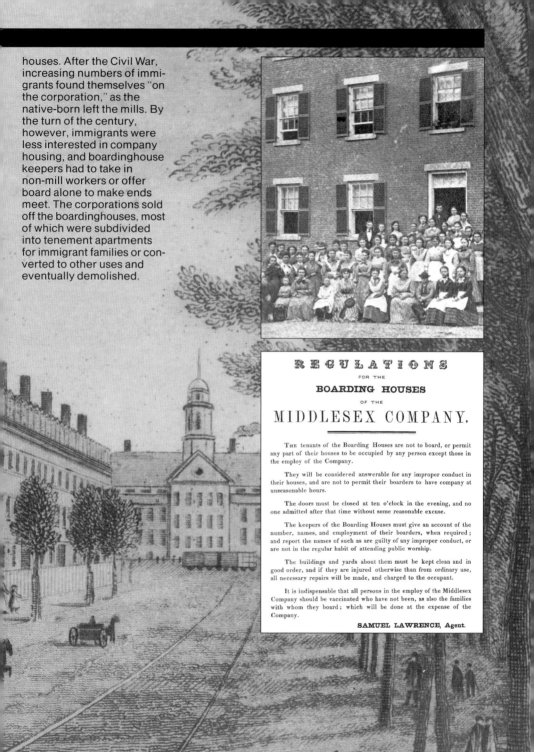

houses. After the Civil War, increasing numbers of immigrants found themselves "on the corporation," as the native-born left the mills. By the turn of the century, however, immigrants were less interested in company housing, and boardinghouse keepers had to take in non-mill workers or offer board alone to make ends meet. The corporations sold off the boardinghouses, most of which were subdivided into tenement apartments for immigrant families or converted to other uses and eventually demolished.

REGULATIONS
FOR THE
BOARDING HOUSES
OF THE
MIDDLESEX COMPANY.

THE tenants of the Boarding Houses are not to board, or permit any part of their houses to be occupied by any person except those in the employ of the Company.

They will be considered answerable for any improper conduct in their houses, and are not to permit their boarders to have company at unseasonable hours.

The doors must be closed at ten o'clock in the evening, and no one admitted after that time without some reasonable excuse.

The keepers of the Boarding Houses must give an account of the number, names, and employment of their boarders, when required; and report the names of such as are guilty of any improper conduct, or are not in the regular habit of attending public worship.

The buildings and yards about them must be kept clean and in good order, and if they are injured otherwise than from ordinary use, all necessary repairs will be made, and charged to the occupant.

It is indispensable that all persons in the employ of the Middlesex Company should be vaccinated who have not been, as also the families with whom they board; which will be done at the expense of the Company.

SAMUEL LAWRENCE, Agent.

Eliza Adams: Mill Worker

In some ways Eliza Adams was a typical "mill girl" who left the New England countryside to work in the mills of Lowell. But she must have been an uncommonly restless soul. She worked at seven different mills or factories, including mills in Pennsylvania and New Jersey, in nine years. In 1841 she left her family's farm in Derry, New Hampshire, when she was 26, to take her first known job at the Lawrence Manufacturing Company in Lowell. Her task, as a "drawing-in girl," was to thread the warp yarns through the reed and harness before weaving began. She was a strong-minded woman, who only a year after she went to work, supported a strike against the owners and wrote a poem calling for worker solidarity.

From the books she left behind we can gather that she shared the spirit of self-improvement found in many of Lowell's female workers. Her seriousness and efficiency, reflected in the account book she started when she was only 19, helped her prosper when she left her last factory job in Chicopee, Massachusetts, to buy a farm in South Hadley. Having never married, she adopted three girls, whom she called "my children" and in whose education she played a guiding role. She died in Derry in 1881.

Eliza Adams' trunk, shown here with some of her personal belongings, saw a good deal of use during her years as an itinerant mill worker. By the time the oval-framed picture was taken, she no longer worked in the mills.

Lowell, Dec. 11 1841

Dear Mother

Arrived in Lowell that evening about dark stopped at №. 30 on the Lawrence James old boarding place. Went to work the next day in the afternoon in №. 5 dress room on the same corporation. I board with James have a good place & find the girls to be a very likly set. Mr. Guild & family with whom we board attend the universalist meeting, he has a shop of dry goods on one of the streets & appears a very good sort of man. I have not seen much of Lowell yet having only 1 half day to walk round. Found Annah Pattee & Selina on the Hamilton. Saw there some double width cloth for sheeting it is something new, will make one wide enough without a seam. If pleasant tomorrow I think of walking out after meeting to see some of the beauties of the place. Last sunday went to the episcopal & saw their performance. My work is drawing in through the harness & reed, the room is very warm, I shall scarcely feel

MERRIMACK RIVER

Immigrant Lowell

Merrimack Mills workers, 1909. By the turn of the 20th century, Lowell was a large, noisy industrial city running on the skills of immigrant workers, including Irish, French Canadian, and Greek. **Facing page:** *Birds-eye view of Lowell, 1876.*

Though Lowell's textile industry continued to expand during the second half of the century, the city lost its dominant position in New England. Between 1850 and 1890 Lowell's textile workforce increased from 10,000 to 15,000. Yet by the end of that period, Fall River, Massachusetts, with more than 19,000 millhands, had displaced Lowell as the nation's largest producer of textiles.

Geographical and technological factors combined to push the city out of its accustomed leading role. As steam power came into general use, freeing mills from the need to locate near rivers, Lowell's inland location on the falls—a crucial advantage in the early Industrial Revolution—hindered its growth. Coastal towns such as Fall River and New Bedford, more accessible to coastal shipping and sea-borne coal, now enjoyed significant transportation advantages over Lowell.

Lowell nevertheless benefited from the new steam technology. After the Civil War, as they approached the limits of the Merrimack River's harnessable power, Lowell firms steadily added steam engines to supplement waterpower. Initially they used steam as a backup for water; later they relied on steam as the primary power source to turn the thousands of new spindles added yearly. By 1870 steam power amounted to about a third of that supplied by Lowell's canal system. By 1880 total steam power had surpassed waterpower.

If Lowell's shift from water to steam power reflected a broader New England pattern, so did its increasing reliance on immigrant labor in the second half of the century. In Lowell's early years the employment of young single women was closely associated in the public mind with the mill city. They had helped to give the rapidly growing industrial complex a good reputation. But Yankee women predominated in the workforce only in the city's first three decades. From mid-century on, successive immigrant groups, beginning with the Irish, took up the lower-paying jobs. After the Irish came French Canadians in the 1860s and 1870s and Greek, Polish, and other nationalities in the 1890s and early 1900s. The onetime Yankee mill town was transformed into a diverse immigrant city.

The Irish had been part of Lowell from the beginning. Kirk Boott employed them to dig canals

As American industry's growing demand for power outran the capacities of waterwheel and turbine, engineers turned increasingly to steam. Unlike water, steam power was not affected by the seasons. From the installation of auxiliary steam power at the mills in Waltham in 1836, the New England textile industry was a leader in the transition to steam.

and raise the first mills. Others found steady work in construction and carting and settled the New Dublin section of the Acre neighborhood. By 1831 about 500 Irish lived in Lowell. Smaller numbers of English and Scottish workers also settled in Lowell, occupying their own distinct neighborhoods, John Bull's Row and Scotch Block. In all, the foreign-born made up roughly 10 percent of Lowell's population in the early years.

When repeated potato blights in Ireland led to what became known as the Famine Migration, Irish numbers in Lowell rose dramatically in the 1840s and 1850s. Although the Irish were finding more work in the mills, they were largely excluded from the mill boardinghouses, which continued to be reserved mainly for the Yankee women. The Irish, too, often preferred to live in their own neighborhoods.

At one firm in Lowell almost 30 percent of the workers in 1850 were Irish. Mill agents, however, still viewed the Irish unfavorably, and would have preferred hiring only Yankees. Initially, they hired the Irish only for the lower-paid jobs in the carding and spinning rooms. But by the outbreak of the Civil War, immigrants were working in the better-paid weaving departments as well.

In its attitude toward the Irish, management reflected a broader prejudice that resulted in frequent confrontations. One of the more serious clashes occurred in 1831, when Yankee laborers attacked the construction site of St. Patrick's Catholic Church and were turned away by rock-heaving Irish. In the 1850s Yankees assailed Catholic schools as "un-American," and an anti-Catholic mob attacked the same church. They were once again repulsed with stones.

Yankee hostility toward the Irish was partly attributable to their effect on the relations between workers and owners. The great numbers of the Irish and their economic need undermined labor protest by Yankee women. Mill management purposely kept Yankees and immigrants apart, and in periods of labor unrest, they played the groups against one another. A large pool of Irish replacement workers had allowed mill agents to disregard the 10-hour campaigns of the 1840s. The Irish also reduced the pressure on mill agents to compete with improving wages in other occupations. The mills in fact reduced wages in this period. By 1860 Lowell's women workers earned less than they had in 1836, despite great increases in productivity.

The Irish presence in the mills brought dramatic change to the Lowell system. Most apparent was the decline in the numbers of single Yankee women and the introduction of family labor. An Irish family typically sent several members, chiefly teenage daughters, to work in the mills. By pooling their low wages the family got by. With this shift from Yankee to Irish labor, the Lowell mills shed themselves of the paternalism that had characterized early Lowell. Mill owners no longer felt obligated to assure potential workers and their parents that the Lowell mills offered an alternative to other New England mills or even to English industrial cities. Compared to the poverty and starvation that the Irish left behind, Lowell must have seemed a marked improvement. To the Yankee women, on the other hand, the changes in Lowell marked the end of an era.

The Civil War hastened the transition from a Yankee to an immigrant workforce in the mills. Believing they could not operate profitably with the anticipated cotton shortages, the Merrimack Company owners shut down that mill. Other companies followed suit, thereby throwing some 10,000 people out of work. (An exception was the wool-producing Middlesex Mills, which prospered from the war by supplying the army with cloth for uniforms.) The Lowell mills sold their stocks of cotton to other New England mills that stayed in operation. When the mills resumed full production after the war, immigrants increasingly took the places formerly occupied by Yankee women.

After the Civil War French Canadians facing economic difficulties in Quebec flocked to the mills. About 600,000 migrated to New England between 1860 and 1900, when they made up about 10 percent of all New Englanders. Their numbers in Lowell increased steadily over these decades, from about 2,000 in 1870 to more than 14,000 by 1905, when they were the largest group of foreign-born in the city. Counting both the immigrant generation and their children, French Canadians made up about 20 percent of Lowell's population on the eve of World War I.

Like the Irish of the 1850s, French Canadians relied upon the earnings of several family members. A study of working-class living standards in Lowell in

Eva and Alvana Desroches worked as weavers in the Boott Cotton Mills. French Canadians came to Lowell in great numbers after 1860— some 14,000 by the time the Desroches sisters had their picture taken in 1903.
Facing page: *1840 depiction of Irish immigrants arriving in America.*

1909 illustrates the workings of the family wage economy within one French-Canadian family. The family consisted of father and mother, six sons between the ages of 2 and 18, and five daughters, ages 8 through 24, all living in a four-room tenement. The father, a blacksmith's laborer, earned on average $8 a week; four of his children combined to bring in another $21 weekly. The food bill for the family came to $12.60 per week, a little under a dollar per person per week or almost half of the family's income. On this income bread dominated the family diet, as they purchased 50 pounds of flour a week, compared to 12 pounds of beef, 4 pounds of pork, and 4 pounds of fish.

Because children helped support the family, parents expected them to subordinate their interests to those of the family. One French Canadian recalled:
My first pay that I ever made was nine dollars and sixty-five cents. A week. . . . I gave my father my envelope and he'd give a dollar for expenses and fifty cents for car fare. That['s] what I had a week. And he kept the rest. It wasn't much, but it helped.

Despite their evident exploitation, French Canadians aroused fear and resentment on the part of Lowell's native-born. A Massachusetts state labor report in 1881 referred to them as "The Chinese of the Eastern States," associating French Canadians with "coolie" (or unfree) labor on the West Coast. The French Canadians lived in a closely-packed neighborhood of wooden tenements known as Little Canada. Recruited by mill agents to help keep down wages, they were blamed by native-born workers for lowering standards within the mills and in the city as a whole. Looked down upon by the earlier Irish arrivals, they founded their own Catholic church and a variety of benevolent and civic organizations. French Canadians responded to the discrimination they faced by looking for strength within their own group.

At the turn of the century, Lowell housed a host of different ethnic groups and nationalities—Irish, French Canadians, Greeks, Poles, Italians, Swedes, Portuguese, Armenians, Lithuanians, Jews, Syrians— over 40 according to school records. By 1908, workers with Yankee parents made up less than 10 percent of all employees in Lowell's mills. Immigrants from southern and eastern Europe predominated among newcomers in Lowell in the years before

The failure of mill owners in early Lowell to accommodate the Irish in company housing set a precedent that significantly influenced community life in the city. Immigrant groups resided away from the mills in their own neighborhoods, where old-world cultures came to terms with the demands of American urban-industrial life. By the turn of the century, Lowell was a microcosm of the broader society—an uneasy blend of many ethnic groups living in distinct neighborhoods.

A 1912 map of Lowell (page 72) showed five major immigrant communities scattered in clusters around the city. Little Canada, which bordered the Northern Canal, had become the primary neighborhood for French Canadians. Greeks concentrated in the Acre along Market Street, while Poles, Portuguese, and Russian Jews had their own enclaves. Within each of these areas, ethnic institutions evolved that catered to the needs of the immigrant group. Ethnic churches were the foremost of these, with a French-Canadian Catholic Church and a Greek Orthodox Church, among others, serving their communities. Parochial schools passed on the native tongue and insured that the American-born generation did not stray too far from the ways of the homeland. Adults also had their cultural institutions: Greek coffee houses and French-Canadian social clubs contributed to the cosmopolitan air of pre-World War I Lowell.

French-Canadian couple, c. 1875

Three generations of Poles, 1913

Greek men relaxing in a coffee house, c. 1914.

Portuguese couple and child, 1905

Young Russian Jewish couple, c. 1910

Syrian bride and groom, c. 1917

By the early 20th century, immigrants typically lived in large, wooden tenement structures called "blocks." In the larger tenements, families lived in apartments of two to four rooms. The blocks, such as those in Little Canada, were crowded, and rooms were often windowless. Even in the better tenements, shared toilets in the hallways or on landings were common. In poorer accommodations, outdoor privies were the rule as late as 1910. The lack of fresh air, light, and baths in many tenements made good hygiene impossible, causing high rates of tuberculosis and infantile intestinal disease.

The passing of company paternalism was one reason for the declining conditions in Lowell. Absentee landlords who bought old mill housing often let the buildings fall into disrepair. As crowded as they had been with Yankee women, sometimes even more immigrants were packed into the same space. Also, as Yankee and second-generation Irish working-class families moved to Lowell suburbs, now connected to the mill districts by trolley, their former homes were often rented out to several Greek or French-Canadian families. The city's growing density magnified all the old problems, and Lowell increasingly suffered from the congestion and health problems typical of early 20th-century industrial cities.

Each immigrant section of Lowell had its distinctive flavor, but the distinctions did not extend to housing. It was usually drab and crowded. Photos on facing page show living conditions in the Acre, 1912 (bottom), and along the Western Canal, 1917. The Irish, first immigrants in Lowell, had once concentrated in the Acre and Chapel Hill neighborhoods. By 1912, the date of the map, other ethnic groups had moved into these neighborhoods and the Irish had dispersed throughout Lowell.

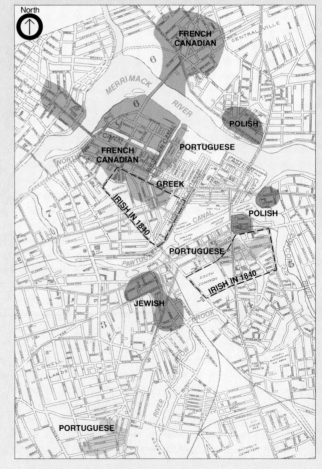

For Irish immigrants, the Catholic Church provided stability and authority in a new world where the living conditions were harsh and the Yankees often hostile. They built a wooden church, St. Patrick's, in 1831, replacing it with a permanent stone structure, above, in 1854.

World War I, with Greece, Austria-Hungary, Portugal, and Poland supplying the largest numbers. The new immigrants settled in distinct neighborhoods, usually crowded blocks of wooden tenements. They suffered under the sometimes harsh conditions that attended the rise of industrial capitalism in America, but also developed a range of creative responses and nurtured their own social organizations.

The largest of these new groups were the Greeks, who concentrated in the Market Street section of the Acre, moving into dilapidated buildings the Irish had left behind. Greeks differed from earlier groups in one important respect. The predominance of single males set them apart from Irish and French-Canadian immigrants. In 1905, after more than a decade of immigration, males outnumbered females among Greeks in Lowell by more than five to one. Young men came to earn money, often with the intention of bringing other family members over later or returning to Greece themselves.

Greeks in Lowell supported many community institutions and owned numerous businesses. A survey in 1912 reported that Greek Lowell had "seven restaurants, twenty coffee-houses, twelve barber shops, two drug stores, six fruit stores, eight shoe-shine parlors, one dry-goods store, four ticket agencies, seven bakeries, four candy stores, [and] twenty-two grocery stores." Ties to the homeland remained strong, and Lowell Greeks continued to identify with political causes in their native land. Churches, a Greek-language school, and a variety of fraternal and cultural organizations, such as the Washington-Acropolis Society, helped maintain their old world culture.

The continuing influx of immigrants led to significant changes to life in Lowell. By the first decade of the 20th century relatively few mill workers were living in company-owned boardinghouses. The corporations sold their boardinghouses, which were often subdivided to house several families. Large numbers of native-born families and second-generation Irish moved to the suburbs to take advantage of moderately-priced housing, now linked to downtown by trolleys. Lowell was a cosmopolitan, but divided, city by the turn of the century.

The flow of immigrants changed the city's politics as well as its population. Before the Civil War, mill agents and overseers had dominated Lowell politics. In one election, managers were accused of pressuring workers to vote for candidates favoring the mills. But when mill managers ended their paternalistic relationship with the workers, they surrendered political control as well. Ethnic groups played an increasingly independent role in Lowell politics in the decades after the Civil War.

Lowell's Irish consistently supported the Democratic Party in the middle decades of the century, and their steadfast support was rewarded in 1882 with the election of the city's first Irish-Catholic mayor. At the turn of the century Republicans and Democrats were evenly matched and both wooed the French Canadians. Yankee reformers in 1911 sought to weaken the hold of Democrats in working-class wards by rewriting the city charter and instituting citywide elections. The reformers won the day but in the next election the Democrats carried the city handily and ushered in an era of Irish-Catholic domination of city politics.

Ethnic divisions marked all phases of Lowell life in the years after 1850. As the Irish, French Canadians, and Greeks developed separate religious, political, and fraternal organizations, ethnicity thwarted united action on the labor front in these decades. Mill agents exploited the divisions by employing immigrant newcomers as strikebreakers. Just as the Irish had entered the mills in large numbers during the unsuccessful 10-hour campaigns of the 1840s, so too French Canadians undermined renewed efforts to reduce the length of the work day in the early 1870s. Greek and Polish millhands in the early 1900s helped to defuse labor unrest in that decade as well. Yet it was often the case that an ethnic group employed as strikebreakers in one generation became a committed part of the labor movement in subsequent generations.

Though women operatives had led Lowell's first strikes, male workers dominated labor protest after the Civil War. In 1875 skilled male mule spinners struck to protest a cut in wages. Management defeated the strike by hiring strikebreakers and by replacing most of the male spinners with women, who tended ring spinning frames requiring less skill to operate. The availability of an alternative spinning technology that supplanted traditional skills and the

The Lawrence Manufacturing Company opened in 1833. It became a knitting mill in the 1890s and continued producing textiles into the 1980s— the longest continuously operating textile company in Lowell.

lack of unity between male craftsmen and female workers doomed the protest.

The participation of organized immigrant workers marked the two major strikes in Lowell before World War I. The 1903 strike, begun by skilled workers represented by the Lowell Textile Council, drew support from French-Canadian, Portuguese, and Polish workers. Though it failed, it did show the need for more cooperation across ethnic and skill lines if future strikes were to succeed.

Lowell mill workers displayed such ethnic unity in a general strike in 1912. Led by organizers from the Industrial Workers of the World (IWW), the Lowell strike followed closely on the heels of the IWW's successful strike in nearby Lawrence. The IWW, the craft unions of skilled workers, and the Greek community all struck together, and the strike ended when management offered a 10-percent wage hike that matched settlements made in other New England textile centers.

The success of the 1912 general strike in Lowell was significant. The strike drew strength from victories in other mill towns, giving Lowell's strikers the confidence they needed to overcome their ethnic divisions. Ironically, the success of New England mill workers in the 1912 strikes must have sent a message to management that contributed to capital flight to the South in the period. The strength of organized workers permitted them on occasion to win concessions from mill owners. The owners, however, had the last word. Those who invested in the Lowell mills were free to disinvest, leaving Lowell workers to face unemployment after the mill closings of the 1920s and 1930s. It is a story as old as capitalism—the movement of capital often leaves misery in its wake.

NOTICE.

The Mule Spinners of Lowell having combined together with a foreign association to coerce their employers to raise their wages, and having made a peremptory demand therefor, and to carry out their purpose having voted to bring out the Lawrence and Massachusetts Corporations on a strike, and those of them employed by these Companies having given notice that they should quit work on the 12th instant.

NOTICE

Is hereby given that if said Spinners shall execute their threat by quitting work accordingly, the services of the Mule Spinners in the employment of this Company will not be required on and after the 14th instant.

MERRIMACK MANUFACTURING Co.,
J. S. LUDLUM, Sup't.
LOWELL, Mass. April 5th 1875

Top: *Della Braga, of Portuguese background, was one of many immigrants who tended Lowell's ring spinning machines. Her photograph was taken at Appleton Mills in 1910.*
Bottom: *When Lowell's mule spinners threatened to strike for higher wages in 1875, management's response was quick and unequivocal: If they struck, their "services [would] not be required."*

By 1900 competitive pressures and technological developments had dramatically changed the working conditions of Lowell millhands. In every department of the mills, fewer workers tended more machinery in 1900 than in 1840. Not only did Lowell operatives tend more machines, but the machinery operated at considerably greater speeds. All told, the demands of textile employment and the toll exacted in terms of workers' health and safety were far greater by 1900 than in the city's early years. A knowledgeable observer in 1903 found that New England mills demanded more work from their operatives than was common even in English mills.

The declining work week compensated somewhat for the quickened pace of work. Still, the mills did not reduce the working hours of their own accord. The hours declined only under steady pressure from state regulation. From an average 73 hours a week in the 1830s and 1840s, a 60-hour week was common by 1874. By 1912 mill owners could demand no more than 54 hours. But that year, when the mills shortened their hours in response to a new state law, management cut daily wages proportionally. This action prompted the famous general strike in Lawrence, led by the Industrial Workers of the World, and successive protests in Lowell, Fall River, and New Bedford. United mill workers prevailed and enjoyed raises rather than the initial pay cuts imposed by management. The unprecedented series of strikes led to important gains for New England's immigrant textile workers.

"Bell Time," by Winslow Homer.

Cotton mule spinners worked barefoot.

IWW's "Big Bill" Haywood (in derby) leads strikers in Lowell, 1912.

The Textile Products of the Lowell Mills

Cotton cloth was always Lowell's major product. But from its earliest years, the mills turned out a variety of textile goods. The Middlesex Company, for instance, manufactured woolen cloth. The Lowell Manufacturing Company was a leading producer of carpets. During the Civil War years, the Lawrence Manufacturing Company moved into the production of hosiery.

Lowell's cotton textiles ranged from pattern weaves and printed cloths to plain "negro cloth," bought by Southern planters to clothe their slaves. The Merrimack Company specialized in calico prints and pioneered in the development of cloth printing technology. Skilled printers were recruited from England in the early years. The head printer hired by the company in 1825 commanded a salary higher than the treasurer's. Other companies specialized in coarse drillings, sheetings, twilled goods, and shirtings, minimizing competition among Lowell textile firms.

78

79

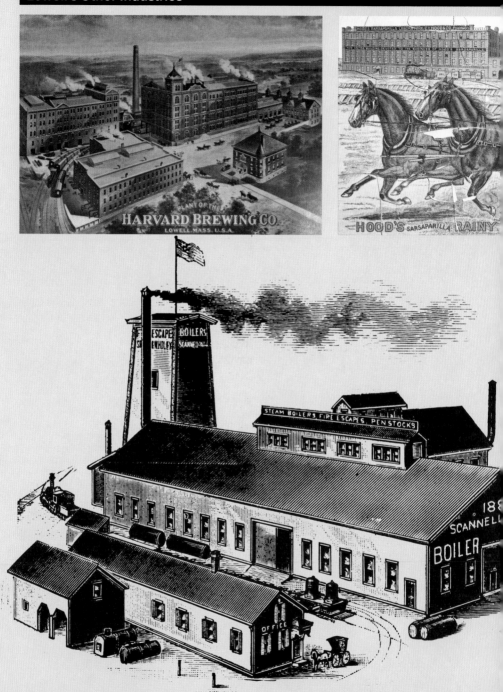

Lowell was dominated by the textile mills in its early years. But throughout the 19th century other important industries grew up in the city. Foremost were textile machinery firms established to meet the demands of textile manufacturers throughout New England. The Lowell Machine Shop and the Kitson Machine Company were the largest of these companies, but there were many others. The Lowell Machine Shop did not limit itself to textile machinery, producing a number of steam locomotives for New England's expanding rail network. Other textile-related firms manufactured and distributed a broad array of mill fixtures, tools, and textile machine parts.

New entrepreneurs built companies unconnected with textiles. A few firms established to supply an expanding national market for patent medicines grew into a major Lowell industry. The Hood and Ayer companies and Father John's Medicine were prominent in this field, pioneering in the skillful use of mass-market advertising. The city's economic base grew more and more diversified: shoe factories, boilerworks, scalemakers, a brewery. During World War I, munitions manufacturers prospered, and the United States Cartridge Company, founded shortly after the Civil War by well-known politician and general Benjamin Butler, was the leading employer in the city.

Signs like this one at the Massachusetts Mills complex were all too common around Lowell after World War I. The city's economic plight deepened as one mill after another shut down, leaving thousands jobless.

Facing page: *Operator tends roving frame, which prepared carded cotton for spinning (c. 1915).*

Decline and Recovery

World War I gave a short-lived boost to Lowell's textile and munitions industries as both profited from large military contracts. As more jobs were created, few could see that the end of Lowell's prosperity was near, or that by 1930 the city's once vital economy would grind to a virtual halt.

There were early signs, but one had to look beyond the production numbers to see them. For several decades after the Civil War, Lowell's textile production had increased steadily, but after 1890 total employment slipped, declining from 17,000 in 1895 to less than 14,000 in 1918. Technological advances made possible gains in output even while mills trimmed their workforce.

Lowell mill owners knew as early as the 1890s that their mills were aging, becoming increasingly non-competitive. Yet mill management chose not to modernize their Lowell operations. They either took their operations elsewhere or used the profits from their Lowell mills to finance modern textile plants in the South.

Southern community and business leaders eager for development actively promoted industrialization by emphasizing the region's advantages of abundant land, cheaper labor, energy sources, lower taxes, and transportation. Promoters also promised New England investors company towns free of union influences and restrictive laws concerning the health and safety of industrial workers. Lowell and other New England mill towns experienced an early version of the capital flight that plagued communities in the northeast and the midwestern industrial heartland in the 1970s and 1980s.

As early as World War I, Lowell firms began to fail or leave town. The Bigelow Carpet Company (formerly Lowell Manufacturing Company, one of the first textile firms in the city) departed in 1914, and Middlesex Mills ceased production in 1918. Other companies took over their plants, but these closings were the first by firms that were part of Lowell's founding almost a century earlier. Then in 1926 came a wave of closings. The Hamilton Company went into receivership, followed by Suffolk, Tremont, and Massachusetts Mills. The Appleton Company moved production to the South, and operations at the Saco-Lowell Shop (formerly the Lowell Machine Shop) shifted north to Maine. By the mid-1930s, of

84

Lowell's first large mills, only the Merrimack, Lawrence, and Boott were still in operation.

The Depression came early to Lowell and stayed. By 1936 total textile employment had dropped to 8,000, only slightly more than it had been a century earlier. Many mills stood empty; others housed a number of small manufacturing firms. Entire mill complexes were demolished, or sections lopped off, to reduce taxes. Parts of Lowell looked like a war-ravaged city.

Families coped as best they could with unemployment during the Depression. One Polish-born worker described how her family survived: "during the summer, dandelion greens were our diet; during the winter we ate hard bread, sweetened with sugar if we were lucky. . . . On rare occasions we would sell something we owned to buy a little meat." Children quit school and took what work they could find. Jobs were scarce, though, and employers often took advantage and made increasing demands on those fortunate enough to be working. Even those with jobs had no assurance of regular work. One former mill worker recalled the Depression:

Many days I walk into the mill, and [the boss] puts his hands up, "No work today." Home you go. They wouldn't tell you anything. You go back the next day, the same thing. The whole week. Wouldn't even tell you if there was no work tomorrow. They waited till you got there.

As the few remaining large mills increased production in the late 1930s, workers responded to the escalating demands made on them. The mills' "stretch-out"—the practice of increasing the workload for the same wage by assigning more machines to workers—recalled a similar demand made on female workers in the 1840s. It had helped drive the Yankee women out of the mills, and it was equally resented by workers a century later. They wanted paid vacations, denied even to those who had been with the company for decades. They asked for improved working conditions, which were hardly better than in the mid-19th century. Yvonne Hoar, who worked in Lowell in the twenties and thirties, recalled what it was like in the Merrimack Mills weave room:

It was the noisiest room you could ever be in. There's machines going and shuttles going back and forth, and sometimes they'd fly off, and they were

The Rebirth of Lowell

By the 1960s Lowell's glory days were far in the past. The city was hard pressed economically, and promising young people were leaving their hometown. Those who stayed were ambivalent about their history, recalling the hard conditions under which they or their parents had worked. With little sense of a worthwhile heritage, many were ready to erase the past and start over. There were even proposals to fill in the city's most distinctive landmarks—its canals—in order to create more downtown real estate.

In the early 1970s a few people with vision, political know-how, and business sense stepped forward with plans to revitalize Lowell. Educator Patrick J. Mogan insisted that any revitalization of the city should be based on its industrial and ethnic heritage. This was the soul of the city—and not incidentally, a key to its economic salvation. Through the efforts of Mogan and others, the city undertook its rehabilitation. The Hu-

man Services Corporation and other community organizations worked with the Lowell Plan, the Lowell Development and Finance Corporation, and business and banking groups in a partnership to guide the city's revival.

After years of study and debate on Mogan's proposal to make Lowell a new kind of national park based on labor and industrial history, Congress established Lowell National Historical Park and the Lowell Historic Preservation Commission in 1978.

Arguing for the park legislation, then Congressman Paul Tsongas, a native of Lowell, defined the idea behind the park:

Twelve years ago Lowell decided that its identity was important. Important to its people and the Nation. There are hundreds of people who should be credited for discovering this America. Many workers . . . wanted the good and the bad of the past preserved, rather than flattened and denied.

The first steps were modest, starting with the renova-

Boott Mills area before and after renovation.

tion of small downtown buildings. The movement quickly gained momentum, benefiting from a new public appreciation for industrial architecture and a belated realization that preservation should embrace working class history and culture. Lowell has once again become a place that is visited by planners from other cities, and even from other countries, who want to follow Lowell's example of using public-private partnerships to bring new life to their communities.

In the late 1950s American readers heard an exuberant new voice. Jack Kerouac (1922-1969) wrote a spontaneous, sometimes raw prose that captured the immediacy of experience. Born of French-Canadian parents in the Centralville area of Lowell, Jean-Louis Kerouac grew up immersed in the city's ethnic, working-class culture. He is best known for his "road" books, such as *Visions of Cody, Dharma Bums*, and especially *On the Road*, which chronicle his restless travels. Through them he became spokesman for what he called the "Beat Generation."

Kerouac (center) at 11 with classmates in "A Musical Sketch."

Kerouac also wrote five books largely set in Lowell, notably *The Town and the City*, in which he calls his hometown "Galloway": *The Merrimac River, broad and placid, flows down to it from the New Hampshire hills, broken at the falls to make frothy havoc on the rocks ... The grownups of Galloway ... work—in factories, in shops and stores and offices, and on the farms all around. The textile factories built in brick, primly towered, solid, are ranged along the river and the canals, and all night the industries hum and shuttle. This is Galloway, milltown in the middle of fields and forests.*

Jack Kerouac smoking shadow, 1953 Lower East Side N.Y., w/ Brakeman's Manual. Allen Ginsberg

pointed things and if they ever hit you, boy, you'd know it. . . . The whole place vibrates. When I come out of there at night I was shaking; I was still in the mill. . . . then they put me up in the finishing room. . . . They were doubling up all the machines so it made that much more work. . . . There we got 13 dollars a week. No matter who you are or where you were in the mill, you got thirteen dollars a week. . . . You really didn't need names, because everyone got thirteen dollars a week. Wouldn't do you any good to complain . . . they were so petrified for their jobs in them days, it was pitiful.

When a union was formed in 1938 to bargain with the Merrimack Mills, women played a significant role in organizing the workers, as their forerunners had a century earlier. After their demands for better wages and working conditions were rejected, they went on strike. Confronted with strikebreakers and called Communists, they had to live on meager strike funds. But far-off events shifted the balance in their favor. War was approaching in Europe and the Federal government was pressuring the mills for cloth. The owners capitulated after seven weeks and the workers returned to the mills.

World War II quickened Lowell's economy. The remaining textile mills in the city—Merrimack, Boott, Ames (the old Lawrence Company), and several others—increased employment dramatically, while the departure of men for military service brought more women into the labor force. The workers could command better wages as other firms with military contracts—Remington, General Electric, and U.S. Rubber—competed with the mills for Lowell workers.

The wartime demand for labor seemed to bring an end to the depression in Lowell that had begun with the mill closings in 1926. Wages shot upward. A typical starting figure of $13 a week in the mills in 1938 rose to $29 by 1943. Earnings in munitions factories were greater still, reaching an average of $37 at Remington for a 48-hour week in 1943.

The boom proved only temporary for Lowell. When the war ended in 1945, orders for munitions and textiles fell off, and the city lapsed into its old economic doldrums. It was clear that the textile industry would not lead Lowell back to prosperity. The city's fortunes were at their lowest in the post-

Jack Kerouac first gained literary notice with the 1950 publication of his autobiographical novel, The Town and the City. *In a style that echoed Walt Whitman and Thomas Wolfe, he told the story of his coming of age in Lowell.*

Portuguese immigrant mill workers pause at their spinning machines for a turn-of-the-century photographer.

war years with the closing of the Boott and Merrimack mills in the 1950s. The latter's mills and boarding-houses soon fell victim to the urban renewal programs of the 1960s, along with the tenement neighborhood of Little Canada. Mill employment all but disappeared, and nothing had yet taken its place. The remaining mill buildings seemed to be bleak reminders of an era of hard work and meager reward. For many residents, remembering the past stirred up feelings of anger and abandonment.

In the 1960s a group of Lowell citizens devised a strategy to revitalize the community, transform the educational system, and stimulate the local economy. Working with urban planners and historians, they laid out a plan for redevelopment based on Lowell's architectural and cultural heritage. Among their proposals was one for a historical park that would present the city as a living museum.

Pragmatic alliances marked this movement from the beginning. Political and business leaders offered support. In 1972 the city council endorsed the idea. Out of this unprecedented cooperation emerged Lowell Heritage State Park in 1974, Lowell National Historical Park in 1978, and the Lowell Historic Preservation Commission. The latter was created in 1978 to assist the park's development, stimulate historic preservation of Lowell's downtown buildings and canals, and develop cultural programs related to the park's themes. Here was a new kind of park, one that mustered the energies, funds, and talents of many groups and government agencies and directed them toward common goals.

Other factors in the 1970s contributed to Lowell's rebirth. The University of Lowell (now the University of Massachusetts Lowell) emerged from the union of Lowell Technological Institute and Lowell State College in 1975. Its mission included support for regional economic development. The arrival in the mid-1970s of Wang Laboratories, then a leader in computers, brought to the city an industry that many hoped would lead to another bright technological future.

The late 1970s and the early 1980s were years of prosperity in Massachusetts, with a soaring economy built around higher education, high technology, and an attractive cultural ambience. In Lowell employment rose as business expanded and over 100 old buildings were rehabilitated and put to new uses. Visitors again came to Lowell, a model of historic preservation and urban revival. But by the late 1980s the boom was over. The region's economy had cooled, the computer industry was tightening its belt, and many companies were closing their doors or relocating elsewhere. Wang found itself challenged by strong competition, requiring it to cut most of its workforce and dramatically restructure its operations.

Boom and bust, technological innovation and obsolescence—these are old themes in Lowell. Many observers see nothing surprising in the current cycle and believe a new mix of technology, improved education, and cultural vitality has positioned the city well for transition into the coming era of internationally interdependent economies.

The city's new pride recalls the spirit of the milltown's boom days. After decades of decline, the population is rising. The most recent immigrants—as essential to industry as their predecessors a century before—come from Cambodia, Laos, Latin America, and other parts of the world. A collection of public art lends interest to the urban scene. Visitors have a multitude of choices: tours of the park, exhibits, festivals, concerts, demonstrations of old skills, a chance to stroll along historic streets. In summer the Merrimack is crowded with boats. The Canalway pedestrian path links the city's waterways with its historic structures.

The citizens of Lowell have made the past a vigorous presence. Historic buildings house new enterprises. Old machinery finds use in new exhibits. Common threads run through the experiences of Lowell's earlier generations of immigrants and those still arriving. If there is any place to observe the beginnings and the development of American industrialization, it is here in Lowell.

The restoration of the Boott Mills clock tower in the 1980s became a symbol of Lowell's new will to remember its industrial heritage.

Part 2

The City as Museum

An 1848 handbook for the Lowell visitor advised "strangers" to obtain a card of introduction from an official at the company he or she planned to visit. No cards are required today, but reservations are necessary for tours and most interpretive programs.

At the visitor center in Market Mills, a multi-image slide show and exhibits explore park themes on Labor, Machines, Capital, Power, and the Industrial City. Visitors can also enjoy art and craft exhibitions, concerts, and city festivals. A nearby outdoor sculpture commemorates working women, fitting in a city whose female workers influenced the course of American labor history.

It surprises some visitors that downtown Lowell resembles a group of islands bordered by natural and artificial waterways. A good way to explore the city is by walking the Canalway, a system of paths bordering the canal system. This path is organized into a downtown inner loop about a mile long and an outer loop that connects with the park trolley and canal boat routes.

The trolleys, modeled on Lowell's turn-of-the-century streetcars, take visitors to the boat landing at Swamp Locks for canal tours. Barges carry visitors through part of the 5.6-mile canal system, where they can see historic gatehouses and locks. Ranger-led tours explore old ethnic neighborhoods, downtown architecture, and other aspects of urban life. Visitors can also take the trolley to several sites for self-guided visits, including the Boott Cotton Mills Museum and the Working People Exhibit at the Patrick J. Mogan Cultural Center.

Spanning the Merrimack Canal across the street from the visitor center is a massive brick vault, once the foundation of the Boston and Lowell Railroad Depot. A restored Boston & Maine locomotive and rail car serve as reminders of the days when this block was the hub of city activity. The corner of Shattuck and Merrimack Streets affords a view of some of the most significant structures in Lowell. Old City Hall, built in 1830, contrasts sharply with the towering current City Hall, completed in 1893. Across the street is St. Anne's Church, built in 1825 by the Merrimack Mills and used by the mill workers. Tree-lined Lucy Larcom Park, between St. Anne's and the Merrimack Canal, is named for a "mill girl" who became a writer and poet. The park is the site

Visitors cruise the Northern Canal along the Great River Wall, the massive granite canal wall rising above the rapids of the Merrimack River.
Facing page: *The weave room at the Boott Cotton Mills Museum.*
Preceding pages: *Market Mills is the front door to Lowell National Historical Park and a good place to begin your visit. The mill complex, part of the old Lowell Manufacturing Company, houses the park visitor center as well as condominiums, a gallery, community video production studios, and a restaurant.*

At the Suffolk Mills Turbine Exhibit, costumed interpreters operate the governor (**top**) *that controls the flow of water into the turbine and a loom* (**bottom**) *powered by the turbine system.*
Facing page: *Vietnamese dragon dancers wind their way through the streets of Lowell in the city's annual folk festival.*

of "INDUSTRY. NOT SERVITUDE!", a multi-piece public art installation which has as its subject the words and ideas of the Lowell Female Labor Reform Association. At the head of the park is the Moody Street Feeder Gatehouse, constructed in 1848 as part of a project that nearly doubled the water supply to power the mills.

Directly in front of the Boott Mills lies Boarding House Park. This outdoor theater offers music and drama celebrating labor and ethnicity. Next to the park is the Patrick J. Mogan Cultural Center. The center is housed in an 1830s Boott corporation boardinghouse block, one of eight that lined the Eastern Canal.

The Boott Cotton Mills Museum traces the industrialization of Lowell. Visitors can tour an early 20th-century-style weave room with up to 100 operating power looms. Other exhibits illustrate Lowell's contribution to the American Industrial Revolution.

The Eastern Canal runs under Bridge Street and curves in front of the Massachusetts Mills. In nearby Eastern Canal Park is the Jack Kerouac Commemorative, a Lowell Public Art site that pays tribute to the Lowell-born author.

Cruises on the canals and the Merrimack River offer fine views of the cityscape and its natural setting. One of the most popular areas in the city is the Vandenberg Esplanade, a 1½-mile landscaped pathway along the north bank of the river. It is a favorite place of joggers, walkers, cyclists, and those who enjoy watching the rowing crews and sailors plying the river.

The park visitor center is located in Market Mills on Market Street. Fees are charged for boat and trolley tours, the Boott Cotton Mills Museum, and some interpretive programs. Reservations are recommended, with special arrangements required for groups. A superintendent, whose address is Lowell National Historical Park, 67 Kirk Street, Lowell, Massachusetts 01852, is in charge. (508) 970-5000.

Lowell's Industrial Heritage Sites

1. Visitor Center
2. Railroad Exhibit
3. New England Quilt Museum
4. Old City Hall
5. Moody Street Feeder Gatehouse
6. St. Anne's Church
7. Agents House/Park Headquarters
8. Mogan Cultural Center/ Boarding House Park
9. Boott Cotton Mills Museum
10. Kerouac Commemorative
11. Lower Locks
12. Industrial Canyon
13. Swamp Locks
14. American Textile History Museum
15. Whistler House Museum of Art
16. Tremont Gatehouse and Power House
17. Suffolk Mills Turbine Exhibit
18. Pawtucket Gatehouse
19. Guard Locks

⊢⊣ Trolley line

● Public art site

UNIVERSITY OF MASSACHUSETTS LOWELL NORTH CAMPUS

Riverside Street

113

Mammoth Road

Veterans

University

of Foreign Wars Highway

Field Hydro Plant

CANAL

Father Morissette

Walkway

NORTHERN

Pawtucket

Avenue

PAWTUCKET FALLS

Northern Canal

18

PAWTUCKET DAM
Elevation 101 feet

Boulevard

Pawtucket

Salem Street

St. Jean Baptiste Church

Moo

Merrim

113

To Lowell Heritage State Park's Vandenberg Esplanade and boathouse

Street

Pawtucket

Wannalancit

FRANCIS GATE PARK

PAWTUCKET CANAL

School Street

Oliver Street

Mt. Vernon Street

Street

Fletcher

NORTH COMMON

St. Patrick's Church

Adams

Street Street

Broadway

19

Francis Gate

UNIVERSITY OF MASSACHUSETTS LOWELL SOUTH CAMPUS

Dutton Street

PAWTUCKET CANAL

CANAL LEVELS

Guard Locks

Swamp Locks
13 foot drop

Lower Locks
17 foot drop

30 foot vertical drop

| UPPER CANAL |
| LOWER CANAL |
| LOWER RIVER |

Middlesex Street

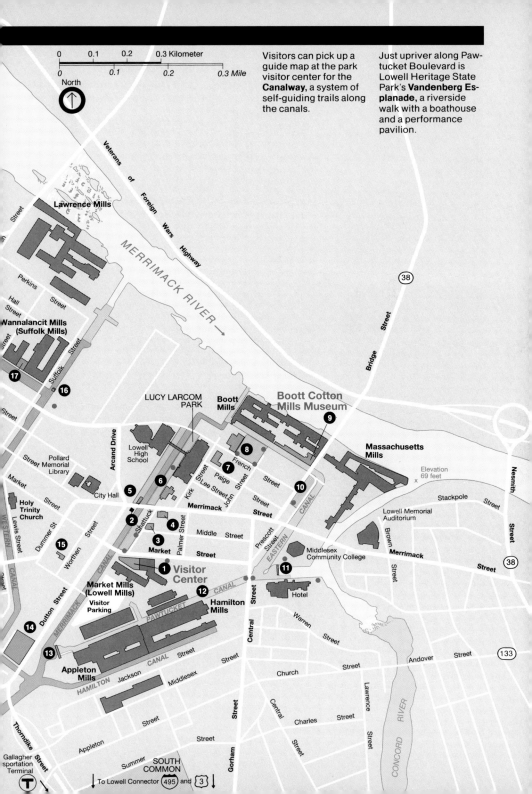

Visitors can pick up a guide map at the park visitor center for the **Canalway,** a system of self-guiding trails along the canals.

Just upriver along Pawtucket Boulevard is Lowell Heritage State Park's **Vandenberg Esplanade,** a riverside walk with a boathouse and a performance pavilion.

0   0.1   0.2   0.3 Kilometer
0   0.1   0.2   0.3 Mile

North

Lawrence Mills

Veterans of Foreign Wars Highway

MERRIMACK RIVER →

Perkins Street

Hall Street

Wannalancit Mills (Suffolk Mills)

Suffolk Street

17

16

LUCY LARCOM PARK

Boott Mills

Boott Cotton Mills Museum

9

Massachusetts Mills

Elevation 69 feet

Arcand Drive

Lowell High School

Pollard Memorial Library

Street

Market Street

Holy Trinity Church

City Hall

5

6

8

7

Paige Street

French Street

Street

Kirk Street

S. Lee Street

John Street

Merrimack   Street

10

EASTERN CANAL

Stackpole Street

Lowell Memorial Auditorium

Nesmith Street

38

Bridge Street

Lewis Street

Dummer St

Worthen Street

15

2

3

Shattuck Street

4

Market Street

Palmer Street

Middle   Street

1   Visitor Center

Prescott Street

Middlesex Community College

11

Brown Street

Merrimack   Street

38

WESTERN CANAL

Dutton Street

Market Mills (Lowell Mills)

Visitor Parking

12

PAWTUCKET CANAL

Hamilton Mills

Hotel

Central Street

Warren Street

MERRIMACK CANAL

14

13

Appleton Mills

HAMILTON CANAL

Jackson   Street

Middlesex   Street

Street

Church   Street

Street

Lawrence   Street

Andover   Street

133

CONCORD RIVER

Thorndike Street

Gallagher Transportation Terminal
T

Appleton   Street

Summer   Street

SOUTH COMMON

Gorham   Street

Central   Street

Charles   Street

Street

↓ To Lowell Connector 495 and 3 ↓

Boott Cotton Mills Museum

A distinctive Lowell landmark is the Boott Mills bell tower, topped with a golden shuttle. The best example of mill architecture in Lowell, Boott Mills is also one of the most significant early industrial sites in the United States.

The first four mills at the Boott were built between 1835 and 1838. Over the next century they were much altered and five more were added. The history of this mill complex parallels the story of Lowell. In the 1840s the Boott Mills produced more than 10 million yards of coarse drillings, fine shirtings, and printed cloth a year. By 1890 more than 2,000 workers operated over 4,000 looms and other textile machines. The workforce dropped to about 700 during the Depression. Government contracts revitalized the mill during World War II. When the Boott closed in 1954, it signalled the demise of large-scale cotton textile manufacturing at Lowell.

The Boott Mills complex was renovated in the 1980s by a private developer with grant assistance from the Lowell Historic Preservation Commission. The countinghouse and Mill No. 6, renovated by the National Park Service, house the Boott Cotton Mills Museum, one of the largest industrial history exhibits in the nation. On the first floor an early 20th-century-style weave room with nearly 100 operating power looms (see page 94) represents the workplace. Upstairs are modern exhibits about the people who made the Lowell factory system work.

In addition to the permanent exhibits, the Boott Gallery features temporary exhibits addressing park themes. The mill building also houses a museum store and historical and educational organizations. The Tsongas Industrial History Center is named for former U.S. Senator Paul Tsongas, a Lowell native who steered through Congress a bill creating the park. A joint project of the National Park Service and the University of Massachusetts Lowell, the center offers teacher training, interactive exhibits, and workshops. The New England Folklife Center of Lowell conserves and supports the rich culture of the region. The National Park Service's Northeast Cultural Resources Center provides technical services in archeology, museum curatorial care, and historic preservation. The Lowell Historical Society holds important collections and promotes local history through publications, exhibits, and other programs.

Amid the din of rows of belt-driven looms in the Boott Weave Room (top), *a costumed mill worker makes a fine adjustment on one of the machines. In 1888 there were over 28,000 looms in the mills of Lowell. Children at the Tsongas Industrial History Center* (bottom) *experiment with the flow of water in a waterpower model.*
Facing page: *The Boott Mills complex by moonlight.*

Market Mills is the gateway to the national park. Although the renovated buildings date from the 1880s and early 1900s, the mill complex was part of the Lowell Manufacturing Company, one of the city's original textile corporations. It now houses the park visitor center, including the multi-image slide show, *Lowell: An Industrial Revelation,* and exhibits, which present the park's major themes: labor, capital, power, machines, and the industrial city.

The Industrial Canyon is a stretch of the lower Pawtucket Canal bordered by the walls of some of Lowell's oldest mills. Visitors walking the Canalway route see a dramatic section of the canal cutting between the Lowell, Appleton, and Hamilton Mills, all founded in the 1820s. Joan Fabrics, Lowell's only active textile manufacturer, occupies a part of the Hamilton complex. In the foreground of this view is sculptor Carlos Dorrien's "Human Construction."

The Pawtucket Gatehouse is the largest such structure in the canal system. It was built between 1846 and 1848 at the entrance to the Northern Canal. Overlooking Pawtucket Dam, the red brick gatehouse controls the flow of water from the river north of the city into the canal. The structure houses ten sluice gates operated by a hydraulic system and a turbine designed by James B. Francis. The system is still active; the Northern Canal feeds a modern hydropower plant downstream.

The Suffolk Mills Turbine Exhibit occupies two levels of the former Suffolk Manufacturing Company, which began producing cotton cloth in 1832. A cutaway section of the floor reveals a restored and functioning 19th-century water turbine. Visitors see a huge pulley wheel, belts, pulleys, and shafting that transfer power to textile machines on the upper floor. Below are a restored 19th-century governor and cavernous work spaces for maintaining the turbines.

The Lower Locks of the Pawtucket Canal is an extensive complex that includes two lock chambers, a dam, gatehouse, and sluice gates that control the flow of water. The massive stone lock chambers were built to help boats navigate the 17-foot drop from the canal to the Concord River. Middlesex Woolen Mills, one of the city's first factories, was located at this site. The area became a symbol of Lowell's revitalization in the 1980s.

Guard Locks on the Pawtucket Canal is best known for the 21-ton wooden gate that has twice saved Lowell from floods. The old Pawtucket Canal was rebuilt as the main feeder canal for the new textile mills. The lock chamber and sluice gates here are the entryway to the system and guarded the canals against high water. Called "Francis' Folly" for engineer James B. Francis, the gate was successfully used in 1852 and 1936 as the rising Merrimack River threatened the city.

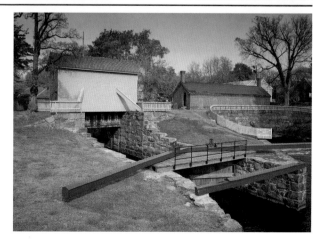

Mogan Cultural Center/Boarding House Park

The Patrick J. Mogan Cultural Center occupies a mostly reconstructed mill workers' boardinghouse block dating from 1837. Named in honor of the educator and planner who conceived the idea of an urban park based on Lowell's culture, the center houses history exhibits, classrooms, a library, and an archive.

Between 1835 and 1838 the Boott Cotton Mills constructed eight rows of boardinghouses across the Eastern Canal from the mills. In the 1980s the Lowell Historic Preservation Commission reconstructed one of the buildings to show the exterior appearance of an original row of boardinghouses. The interior is a fitting place for an exhibit on "The Working People: Mill Girls, Immigrants, and Labor."

Visitors walk into a boardinghouse furnished in the style of the 1850s. The first floor includes a dining room, kitchen, and keeper's room. Upstairs is a bedroom representing the cramped quarters in which the women lived. The area is brought to life by an audio program that lets visitors eavesdrop on the conversations of women workers.

From the "mill girl" area, visitors pass into an exhibit area that looks at the social world of the immigrant groups in Lowell since the 1820s. A timeline marks the arrival of each group, from the first Irish laborers to recent Cambodian refugees. Maps indicate countries of origin, and immigrants or their descendants describe their experiences in video programs. A space for changing exhibits provides community groups the opportunity to present their own stories.

The facility also houses the Center for Lowell History of the University of Massachusetts Lowell—a library and archive of Lowell photographs, books, and other documents—and the University's downtown Continuing Education Center.

Boarding House Park is the front lawn of the center. The one-acre park steps down towards the Eastern Canal, where the park trolley stops. Shade trees and broad grassy areas make the park a popular lunch spot, and its outdoor stage is the site of concerts, plays, and festivals.

In the 1840s four women usually occupied a bedroom of this size **(top).** *The experience of the mills' later workers is depicted in the immigrants section of the exhibit* **(bottom).** *Father John's, one of Lowell's many popular patent medicines, was named after an Irish immigrant priest.* **Facing page:** *The reconstructed Boott Mills boardinghouse (Mogan Cultural Center).*

Old City Hall was Lowell's first town hall, but it bears little resemblance to its original appearance in 1830. The exterior has been preserved in the Georgian Revival style of 1896, reflecting alterations made at the time to enhance its use as a commercial building. A number of distinguished Americans, including Daniel Webster and Abraham Lincoln, spoke in this hall.

St. Anne's Church and Rectory are among the oldest standing structures built by Lowell's corporate founders. In 1825 the Merrimack Manufacturing Company constructed the church for its workers and the town's new settlers. The Gothic Revival style church and Greek Revival parsonage were built of stone excavated during the construction of the adjoining canal. The "mill girls," who were required to attend services, had pew rent deducted from their wages in the early years.

The Agents House, built between 1845 and 1847, is one of the best examples of early corporate housing in Lowell. This 2½-story brick duplex was a residence for the agents of the Massachusetts and Boott Cotton Mills until 1901. It was later used as a rooming house, school offices, and a medical clinic. Park headquarters is housed here. Plans call for exhibits about mill management and the growth of this industrial city.

The American Textile History Museum is located in Lowell's old Kitson Machine Shop. The museum houses artifacts, books, and images chronicling the story of cloth-making in America. Exhibits include an 1860s woolen mill and a functional 1950s weave room. Built in 1866, the Kitson Machine Shop manufactured textile machinery. A Kitson picker, used to clean cotton fibers prior to carding, is on display. The museum includes the Textile Conservation Center, a research library, a museum shop, and cafe.

The Whistler House Museum of Art is the birthplace of American artist James Abbott McNeill Whistler. His father, George Washington Whistler, was Chief Engineer of the Locks & Canals Company, which in 1823 began using the house as a residence for its chief engineers. James was born there in 1834. In 1908 the Lowell Art Association opened the house as an art center and home for its collection, which includes works by Whistler, John Singer Sargent, and many noted regional artists.

The Massachusetts Cotton Mills, incorporated in 1839, ceased operating almost a century later and was renovated in the late 1980s as a housing complex. The green space of nearby Eastern Canal Park includes the **Jack Kerouac Commemorative.** This outdoor art honors the Lowell native who wrote more than 20 books, including *On the Road* and five novels based on his youth in Lowell. Excerpts from his writings are inscribed on eight polished granite columns.

The silhouette of Francis Cabot Lowell is combined with a thread spool/beehive form symbolizing industry. Massive post and lintel blocks flanking a bridge express the internal forces necessary to build anything: a frame, a home, a city. A bronze figurative sculpture is a tribute to the women of Lowell. Robert Cumming's "The Lowell Sculptures," (right) Carlos Dorrien's "Human Construction," (above) and Mico Kaufman's "Homage to Women" (above right) are part of the Lowell Public Art Collection.

The collection comprises artworks that address the themes of the American Industrial Revolution in an

urban setting. These pieces, integrated into small parks, plazas, and structures along the Canalway, generate dialogue, self-examination, and celebration of Lowell's ethnic and industrial traditions.

Other works around the city include a bronze sculpture of a canal builder, "The Worker," (right) by Ivan and Elliot Schwartz; Dimitri Hadzi's "agápetimé"; "The Big Wheel," an original pulley wheel from the Lawrence Mills; Michio Ihara's "Pawtucket Prism"; and the "Jack Kerouac Commemorative," (pg. 107) by Ben Woitena with Brown and Rowe Landscape Architects. The newest works include the bronze "Stele for the Merrimack" by Peter Gourfain, with images of the people and the flora and fauna that depended on the Merrimack River; Michael Singer's "Pawtucket Falls Island Project," which integrates sculpted stone with growing plants; and Ellen Rothenberg's "INDUSTRY, NOT SERVITUDE!", a sculptural series representing the struggles of the 19th-century working women of Lowell.

Numbers in italics refer to photographs, illustrations, or maps.

Facing page: *The park trolley stops in front of the Boott Cotton Mills Museum.*

National Park Service

For Further Reading The best general histories of Lowell are *Cotton Was King: A History of Lowell, Massachusetts,* ed. Arthur L. Eno, Jr., 1976, and *The Continuing Revolution,* ed. Robert Weible, 1991. For an overview of the Merrimack Valley region, see *The Valley and Its People: An Illustrated History of the Lower Merrimack,* by Paul Hudon, 1982. For an environmental history approach to Lowell, see Theodore Steinberg's *Nature Incorporated: Industrialization and the Waters of New England,* 1991. A good account of the New England textile industry is *The Run of the Mill,* by Steve Dunwell, 1978.

Lowell's labor history is explored in two volumes compiled by Mary H. Blewett: *Surviving Hard Times: The Working People of Lowell,* 1982, and *The Last Generation: Work and Life in the Textile Mills of Lowell, Massachusetts, 1910-1960,* 1990.

For a closer look at the "mill girl" era, see *Women at Work: The Transformation of Work and Community in Lowell, Massachusetts, 1826-1860,* 1993, and *Farm to Factory: Women's Letters, 1830-1860,* 1993, by Thomas Dublin, and Harriet Robinson's *Loom and Spindle or Life Among the Early Mill Girls,* 1898, reprinted 1976.

On the evolution of Lowell as an industrial city, see *Enterprising Elite: The Boston Associates and the World They Made,* by Robert Dalzell, 1987, and *Nathan Appleton: Merchant and Entrepreneur, 1779-1861,* by Frances W. Gregory, 1975. For a critical history of an original Lowell textile company, see *The Course of Industrial Decline: The Boott Cotton Mills of Lowell, Massachusetts, 1835-1955,* by Laurence F. Gross, 1993.

For information on immigrant Lowell, see *The Paddy Camps: The Irish of Lowell, 1821-61,* by Brian Mitchell, 1988, and *Immigrant Odyssey: A French-Canadian Habitant in New England,* by Felix Albert, translated by Arthur L. Eno, Jr., ed. Frances Early, 1991.

Related Industrial Sites Manchester, N.H., is the site of Amoskeag Mills, once the largest cotton factory in the world. Exhibits and historical tours are available at the Manchester Historic Association museum. In Lawrence, Mass., there are programs and exhibits at Lawrence Heritage State Park and Immigrant City Archives. The Charles River Museum of Industry at Waltham, housed in the old Boston Manufacturing Co. mill complex, features exhibits on major industries along the Charles.

By 1880 Fall River, Mass., was the nation's preeminent textile center. Fall River Heritage State Park interprets the city's industrial and maritime past.

The Slater Mill Historic Site in Pawtucket, R.I., is a National Historic Landmark. The first successful textile factory in the U.S., Slater Mill began producing yarn with water-powered carding and spinning frames in 1790. Blackstone River Valley National Heritage Corridor links Rhode Island sites with Massachusetts mills on the Blackstone River and Canal.